Steffen Stern

Dynamical dark energy and variation of fundamental "constants"

Steffen Stern

Dynamical dark energy and variation of fundamental "constants"

A study of experimental probes and theoretical models

Südwestdeutscher Verlag für Hochschulschriften

Impressum/Imprint (nur für Deutschland/ only for Germany)
Bibliografische Information der Deutschen Nationalbibliothek: Die Deutsche Nationalbibliothek verzeichnet diese Publikation in der Deutschen Nationalbibliografie; detaillierte bibliografische Daten sind im Internet über http://dnb.d-nb.de abrufbar.
Alle in diesem Buch genannten Marken und Produktnamen unterliegen warenzeichen-, marken- oder patentrechtlichem Schutz bzw. sind Warenzeichen oder eingetragene Warenzeichen der jeweiligen Inhaber. Die Wiedergabe von Marken, Produktnamen, Gebrauchsnamen, Handelsnamen, Warenbezeichnungen u.s.w. in diesem Werk berechtigt auch ohne besondere Kennzeichnung nicht zu der Annahme, dass solche Namen im Sinne der Warenzeichen- und Markenschutzgesetzgebung als frei zu betrachten wären und daher von jedermann benutzt werden dürften.

Verlag: Südwestdeutscher Verlag für Hochschulschriften Aktiengesellschaft & Co. KG
Dudweiler Landstr. 99, 66123 Saarbrücken, Deutschland
Telefon +49 681 37 20 271-1, Telefax +49 681 37 20 271-0, Email: info@svh-verlag.de
Zugl.: University of Heidelberg, Diss., 2008

Herstellung in Deutschland:
Schaltungsdienst Lange o.H.G., Berlin
Books on Demand GmbH, Norderstedt
Reha GmbH, Saarbrücken
Amazon Distribution GmbH, Leipzig
ISBN: 978-3-8381-0254-2

Imprint (only for USA, GB)
Bibliographic information published by the Deutsche Nationalbibliothek: The Deutsche Nationalbibliothek lists this publication in the Deutsche Nationalbibliografie; detailed bibliographic data are available in the Internet at http://dnb.d-nb.de.
Any brand names and product names mentioned in this book are subject to trademark, brand or patent protection and are trademarks or registered trademarks of their respective holders. The use of brand names, product names, common names, trade names, product descriptions etc. even without a particular marking in this works is in no way to be construed to mean that such names may be regarded as unrestricted in respect of trademark and brand protection legislation and could thus be used by anyone.

Publisher:
Südwestdeutscher Verlag für Hochschulschriften Aktiengesellschaft & Co. KG
Dudweiler Landstr. 99, 66123 Saarbrücken, Germany
Phone +49 681 37 20 271-1, Fax +49 681 37 20 271-0, Email: info@svh-verlag.de

Copyright © 2009 by the author and Südwestdeutscher Verlag für Hochschulschriften Aktiengesellschaft & Co. KG and licensors
All rights reserved. Saarbrücken 2009

Printed in the U.S.A.
Printed in the U.K. by (see last page)
ISBN: 978-3-8381-0254-2

Dynamical dark energy and variation of fundamental "constants"

In this thesis we study the influence of a possible variation of fundamental "constants" on the process of Big Bang Nucleosynthesis (BBN). Our findings are combined with further studies on variations of constants in other physical processes to constrain models of grand unification (GUT) and quintessence. We will find that the ^{7}Li problem of BBN can be ameliorated if one allows for varying constants, where especially varying light quark masses show a strong influence. Furthermore, we show that recent studies of varying constants are in contradiction with each other and BBN in the framework of six exemplary GUT scenarios, if one assumes monotonic variation with time. We conclude that there is strong tension between recent claims of varying constants, hence either some claims have to be revised, or there are much more sophisticated GUT relations (and/or non-monotonic variations) realized in nature. The methods introduced in this thesis prove to be powerful tools to probe regimes well beyond the Standard Model of particle physics or the concordance model of cosmology, which are currently inaccessible by experiments. Once the first irrefutable proofs of varying constants are available, our method will allow for probing the consistency of models beyond the standard theories like GUT or quintessence and also the compatibility between these models.

Contents

I Introduction and prerequisites   1

1 Introduction   2

2 Variation of "constants"   4
- 2.1 The laws of physics and the constants of nature 4
- 2.2 The question of constancy 4
- 2.3 Theoretical arguments for variation of constants 5
- 2.4 Equivalence principles and possible violations 6
- 2.5 Variation of dimensionful parameters 7
  - 2.5.1 The chiral limit 7
- 2.6 Probes of varying constants 7
- 2.7 Fine-tuning of constants and the anthropic principle 8

3 Cosmology   9
- 3.1 General relativity and the basics of cosmology 9
  - 3.1.1 General relativity 9
  - 3.1.2 The basics of cosmology 9
- 3.2 The concordance model 12
  - 3.2.1 Historical development 12
  - 3.2.2 Our current picture of the Universe 12
- 3.3 Cosmological parameter values 14

4 The Standard Model and beyond   16
- 4.1 The Standard Model of particle physics 16
- 4.2 Running of couplings 17
- 4.3 The necessity of a "theory beyond" 19
- 4.4 Supersymmetry and the MSSM 20
- 4.5 Grand unification 20
- 4.6 Variations in a GUT framework 22
  - 4.6.1 Variation of the electromagnetic coupling 23
  - 4.6.2 Variation of the QCD scale 24
  - 4.6.3 Conversion of units 24

5 Models of quintessence   25
- 5.1 Problems of the cosmological constant 25
- 5.2 Basics of quintessence 25
- 5.3 Crossover quintessence models 27
- 5.4 Growing neutrino mass models 30

	5.4.1 Stopping growing neutrino model	31
	5.4.2 Scaling growing neutrino model	33
5.5	A short note on string theory	34

II Big Bang Nucleosynthesis — 35

6 Big Bang Nucleosynthesis — 36
- 6.1 Why BBN? ... 36
- 6.2 How will we study BBN? ... 36
- 6.3 The process of BBN ... 37
- 6.4 The physics of BBN ... 39
  - 6.4.1 Cosmological background equations ... 39
  - 6.4.2 Initial conditions ... 40
  - 6.4.3 The element synthesis process ... 41
- 6.5 Nuclear reaction rates and the Q value ... 42
- 6.6 Nuclear reactions important for BBN ... 43
  - 6.6.1 The $n \leftrightarrow p$ reaction rate ... 44
  - 6.6.2 The $n + p \to D + \gamma$ reaction rate ... 45
  - 6.6.3 Charged particle reaction rates ... 45
- 6.7 The simulation of the BBN process ... 47
  - 6.7.1 Numerical aspects of the BBN simulation ... 48
- 6.8 Observational situation and uncertainties ... 49
  - 6.8.1 ^{4}He ... 49
  - 6.8.2 Deuterium ... 49
  - 6.8.3 ^{3}He ... 50
  - 6.8.4 ^{7}Li ... 52
  - 6.8.5 ^{6}Li ... 53
  - 6.8.6 Theoretical predictions ... 53

7 BBN with varying constants — 54
- 7.1 Nuclear and fundamental parameters ... 54
- 7.2 Nuclear parameters relevant for BBN ... 54
- 7.3 Nuclear parameter dependence ... 55

8 From nuclear to fundamental parameters — 57
- 8.1 From nuclear to fundamental parameters ... 57
  - 8.1.1 Pion mass ... 58
  - 8.1.2 Neutron and proton mass ... 59
  - 8.1.3 Neutron lifetime ... 60
  - 8.1.4 Binding energies ... 61
- 8.2 The response matrices ... 63
- 8.3 Comparison to other studies ... 64

9 Constraints on variations — 65
- 9.1 Bounds on separate variations of fundamental couplings ... 65
- 9.2 Variations of abundances in unified models ... 65
  - 9.2.1 Linear results ... 67
  - 9.2.2 Nonlinear results ... 67

III Unifying cosmological and late-time variations 71

10 Experimental tests of variations 72
- 10.1 BBN 72
- 10.2 CMB 73
  - 10.2.1 Effect of "varying constants" at CMB and η 74
- 10.3 Quasar absorption spectra 74
- 10.4 The Oklo natural reactor 77
- 10.5 Meteorite dating 78
- 10.6 Bounds on the variation of $G_{\rm N}$ 78
- 10.7 Atomic clocks 80

11 Variations from BBN to today in GUTs 82
- 11.1 GUT relations 82
- 11.2 Variations in six different unified scenarios 84
  - 11.2.1 Varying α alone 85
  - 11.2.2 Scenario 1: Varying gravitational coupling 85
  - 11.2.3 Scenario 2: Varying unified coupling 87
  - 11.2.4 Scenario 3: Varying Fermi scale 87
  - 11.2.5 Scenario 4: Varying Fermi scale and SUSY-breaking scale . . . 87
  - 11.2.6 Scenario 5: Varying unified coupling and Fermi scale 88
  - 11.2.7 Scenario 6: Varying unified coupling and Fermi scale with SUSY 91
- 11.3 Epochs and evolution factors 92
  - 11.3.1 Epochs 92
  - 11.3.2 Evolution factors 93
  - 11.3.3 Monotonic evolution with unification 94
  - 11.3.4 Tension between the ^{7}Li problem and variation of μ 96

12 Probing quintessence models 98
- 12.1 Crossover quintessence 98
- 12.2 Models with growing neutrinos 100
  - 12.2.1 The stopping growing neutrino model 100
  - 12.2.2 Global fit to the scaling growing neutrino model 101
- 12.3 Tests of the weak equivalence principle 102
- 12.4 Bounds on present-day variation 103

13 Conclusion and outlook 106

Acknowledgements 108

A Conventions 109
- A.1 Symbols and abbreviations 109

List of tables 111

List of figures 112

Bibliography 113

Part I

Introduction and prerequisites

Chapter 1

Introduction

The constants of nature

Since the time of Newton, the constancy of the fundamental laws of nature has been undoubted. Comparing and reproducing experiments have been at the root of the scientific approach: A physical experiment which we perform today will have the same outcome as the same experiment performed tomorrow[1]. Neglecting local gravitational effects, it should also not matter where we perform the experiment. Hence, it has been unquestionable for a long time that the laws of nature are constant over space and time. Moreover, Einstein formulated this space- and time independence of physics in his strong equivalence principle, making it an essential part of his theory of general relativity.

Today's view of this question is somewhat different, at least from theoretical aspects. Even though compelling evidence for changes in the laws of physics has up to now not been found, we have to admit that we are still lacking a profound test of this constancy. In the past, the laws of physics have only been thoroughly tested on time and length scales accessible by mankind, *i.e.* on timescales of years and on length scales that do not go beyond the size of our solar system[2]. Only recently astrophysics and cosmology have opened a door to test physics on immensely broader scales, reaching out to unimaginable length scales of several gigaparsecs and going back in time to the very beginning of our Universe.

This thesis will deal with probes of possible variations of constants throughout the whole accessible history of the Universe. In a first part, we will study one of the most distant (in time and space) events where physics can be applied and tested, primordial nucleosynthesis. It is the process during which the light elements of our Universe were formed and which happened when our Universe was only one minute old, extremely hot and dense. If physics was really subject to variations, primordial nucleosynthesis is a prime candidate for any studies of this kind. In a second step the obtained results will be combined together with further tests of varying constants at later times to derive a "history of variations". Finally, we will show how these results can be used to test models beyond standard physics which currently cannot be accessed directly by experiments.

---

[1] Neglecting experiments which incorporate probabilities, for instance quantum mechanical effects.
[2] Note that general relativity has furthermore not been tested on length scales smaller than about 1mm.

Outline

In this thesis I will work out the influence of varying "constants" on the process of primordial nucleosynthesis and implied constraints to theories beyond standard physics. In the next chapter, I will give a short introduction to variations of physical constants, some historical remarks and theoretical motivations. Chapter 3 will lay the theoretical framework for our understanding of the Universe as a whole, explaining general relativity and the main laws of cosmology. In Chapter 4 I will introduce the concepts of supersymmetry and grand unified theories (GUTs) which are widely accepted as extensions of the Standard Model of particle physics. Chapter 5 will introduce quintessence models which can yield variations of constants.

Part II will focus on the details of one particular process in the history of our Universe, Big Bang Nucleosynthesis (BBN). Chapter 6 will explain the standard process of BBN and the physics behind. Chapter 7 will introduce the possibility of varying constants in the process of BBN, and Chapter 8 will demonstrate how one can relate the results to variations of the Standard Model parameters. Finally, in Chapter 9, the observed element abundances will be used to derive constraints on variations of fundamental parameters.

In Part III I will study relevant tests of varying constants from the Big Bang until today. Chapter 10 gives an overview over tests of varying constants, and in Chapter 11 I will combine these tests within six different GUT models, showing how variations of constants can in principle be used to probe models of grand unification. Using the six GUT models, Chapter 12 shows how models of quintessence can be probed under the assumption of grand unification.

Finally, in Chapter 13, I will sum up the findings of this thesis and give some final conclusions and outlook.

The work on this thesis has led to four main publications:

- Michael Doran, Steffen Stern, Eduard Thommes,
  Baryon Acoustic Oscillations and Dynamical Dark Energy,
  JCAP 0704:015 (2007) [DST06] (not in focus of this thesis)

- Thomas Dent, Steffen Stern, Christof Wetterich,
  Primordial nucleosynthesis as a probe of fundamental physics parameters,
  Phys. Rev. D 76, 063513 (2007) [DSW07]

- Thomas Dent, Steffen Stern, Christof Wetterich,
  Unifying cosmological and recent time variations of fundamental couplings,
  Phys. Rev. D 78, 103518 (2008) [DSW08.1]

- Thomas Dent, Steffen Stern, Christof Wetterich,
  Time variation of fundamental couplings and dynamical dark energy,
  Preprint arXiv:0809.4628, accepted by JCAP [DSW08.2]

Chapter 2

Variation of "constants"

2.1 The laws of physics and the constants of nature

The fundamental laws of physics, represented by the Standard Model of particle physics and Einstein's theory of general relativity, consist of two parts. One part is the mathematical form of the laws (*e.g.* the $1/r$ behavior of Newton's theory of gravity), the other part is the actual strength of the interactions relative to each other. Whilst the first part, the mathematical form of the laws of nature, can be derived from considerations of fundamental symmetries of nature[1], the second part has to be put into the theories "by hand" in form of about 27 - from a theory standpoint a priori absolutely arbitrary - numerical values, the constants of nature. Tab. 2.1 gives a list of these fundamental constants[2] for the Standard Model of particle physics and general relativity[3]. Note that the list of fundamental parameters gets much larger when going to theories beyond the Standard Model, *e.g.* supersymmetry (see Sec. 4.4). Up to now it is unclear where these constants come from and if they are "real" constants in the sense that their numerical values are fixed once and for all.

2.2 The question of constancy

The question if the constants of nature are actually constant was probably first raised by Dirac [Dirac37, Dirac38, Dirac79]. In his "large numbers hypothesis", he argues that very large (or small) dimensionless constants must not enter in basic laws of physics. Based on his numerological principle, he suggests that very large numbers rather characterize the state of the Universe, specifically the time which has passed since the Big Bang. For instance, he finds that the age of the Universe in atomic time, $H_0 e^2 / m_e c^2 \approx 2 \times 10^{-41}$, is of the same order of magnitude as the ratio of electrostatic to gravitational force between proton and electron, $G_N m_p m_e / \frac{e^2}{4\pi\epsilon_0} \approx$

---

[1] For example, the Standard Model of particle physics is obtained when demanding a local $SU(3) \times SU(2) \times U(1)$ symmetry. See Sec. 4.1.

[2] As will be explained in Sec. 2.5, only ratios of masses are measurable fundamental parameters. Hence, in fact one can get rid of one the mass terms in Tab. 2.1, for instance by defining all masses with respect to the Planck mass. This would reduce the number of fundamental parameters by one.

[3] In cosmology some more free parameters turn up which have to be determined by observations, for instance those describing the composition of our Universe. However, it is assumed that these parameters can in principle be obtained from some fundamental laws of physics once the processes in the very early stage of the Universe are better understood.

Type of constant	Number of parameters
3 coupling constants	3
masses of 6 quarks	6
CKM matrix (3 angles + 1 complex phase)	4
masses of 3 leptons	3
Higgs mechanism	2
strong CP phase	1
masses of 3 neutrinos	3
PMNS mixing matrix for neutrinos	4
gravitational constant	1
in summa	27

Table 2.1: The fundamental constants of nature.

4×10^{-40}. Consequently, he suggests that also the latter quantity should vary with cosmic time. Attributing the variation to the gravitational sector, the intensity of all gravitational effects would then decrease with a rate of about 10^{-10} y^{-1}. It was quickly found that this would lead to astrophysical effects [Chandrasekhar37] which could not be detected in the following time. Hence, Dirac's theory was finally abandoned, but the discussion on varying constants had started[4].

In 1961, Brans and Dicke [BransDicke61] used Mach's principle[5] to derive what we now call a "scalar-tensor theory". In their model, the gravitational constant is replaced by a scalar field which can vary in space and time. Besides others, models of this kind are still being considered as theoretical arguments for variation of constants.

2.3 Theoretical arguments for variation of constants

In high-energy theories such as string theory, which unifies gravity with the Standard Model of particle physics, our low-energy laws of physics appear as an effective theory whose parameters are set dynamically by vacuum expectation values which break the "higher" symmetry. In particular string theory offers a plethora of possibilities to introduce variations of constants, for instance due to the fact that it is formulated with 10 (or 11) spacetime dimensions which need to be compactified in order to arrive at the 4 spacetime dimensions of the Standard Model (we will give some more details in Sec. 5.5). Similar considerations also apply to other theories with extra dimensions, for instance the possibility of varying constants in Kaluza-Klein theories has been studied in [Marciano83]. Hence, both temporal and spatial variations of constants are from a theoretical standpoint well founded, even though those high-energy theories mostly do not give any hint on the actual size of the variations.

Also, "low-energy" theories, for instance theories which extend the concordance model of cosmology by introducing a cosmological scalar field, allow variations of constants. In this thesis we will concentrate on theories of coupled quintessence in which constants can depend on cosmic time and the environment.

This thesis will examine the possibility of variations of constants from today back

---
[4]See for instance [Uzan02] for a more complete review of the history of varying constant theories.
[5]There are different formulations of Mach's principle. In Brans' and Dicke's argument it states [Brans05] that the gravitational constant should be a function of the mass distribution in the universe.

to the time of Big Bang Nucleosynthesis (BBN). During BBN, the composition of the Universe was quite different from today's composition (concerning temperature and pressure). Hence, composition dependent effects which might cause spatial variations today might have caused variations at BBN time. However, since the Universe was almost homogeneous at BBN, these variations can effectively be treated as a time-dependent effect. This thesis will not evoke the question of space-dependence of constants but treat possible variations at BBN as purely temporal effects.

2.4 Equivalence principles and possible violations

Particle theory is based on Poincaré covariance. In quantum field theory (QFT), we demand that each of the fundamental fields is a representation of the Poincaré group. Hence, amongst others, invariance under spacetime translations is automatically built in. However, we can still implement spacetime variations by introducing additional dynamical fields, whose values are determined by the fields' own actions and their couplings to the rest of the theory. While the theory as a whole remains Poincaré invariant, variations in measurable quantities can still arise if the solution for the additional fields has a nontrivial spacetime dependence.

This discussion can be extended to general relativity (GR), which is also based on symmetry principles that are apparently violated by variations of constants. In particular, GR is based on the strong equivalence principle, which can be decomposed into the following symmetries

- Weak equivalence principle: The trajectory of a freely falling test body only depends on its initial position and velocity and is independent of its composition.

- Local Lorentz invariance: The outcomes of any experiments (whether gravitationally or not) in a laboratory moving in an inertial frame of reference are independent of the velocity of the laboratory.

- Local position invariance: Outcomes of experiments (whether gravitationally or not) do not depend on their position in space and time.

A space or time variation of fundamental constants obviously violates local position invariance. Also, as the gradient of any varying fundamental parameter defines a direction in spacetime, local Lorentz invariance is violated. Finally, it has been shown (see *e.g.* [Nordtvedt02]) that any space-time variation of fundamental constants will necessarily lead to an additional gravitational force, hence also the weak equivalence principle will be violated[6]. As probes of general relativity so far do not find any violation of the theory, we can immediately conclude that variations of constants must be extremely tiny. Note, however, that GR has only been tested on relatively small time scales and also only on length scales from 1mm to the size of our solar system.

---

[6]We will work out the relation between violation of the weak equivalence principle and variation of constants in Sec. 12.3.

2.5 Variation of dimensionful parameters

When measuring or estimating possible variations of constants, one always has to keep in mind that the variation of any dimensionful quantity is not physically well-defined, as one always has to specify how the dimension (*e.g.* the unit [Energy]) is defined. In general, a dimensionful quantity can only be measured by comparison with another dimensionful quantity, so in fact only dimensionless ratios are measurable. For example, measurements of variations of the electron mass m_e are only well-defined if one states how the mass unit is defined, for instance by choosing a system of units where the mass scale is kept fixed. Popular system of units are the "Einstein frame" where the Planck mass M_P is kept constant, or the "Jordan frame" where some particular particle physics scale is kept fixed. Considering the electron mass in the Einstein frame, the actually measured varying quantity (without system of units ambiguities) is then rather m_e/M_P.

In the part of this thesis which deals with Big Bang Nucleosynthesis, we use a system of units where the QCD invariant scale Λ_{QCD} is kept constant. This is convenient for dealing with nuclear reactions, where the energy scales are mainly determined by the strong interaction. Thus the variations of dimensionful parameters include implicitly some power of Λ_{QCD}. For example, if we take the electron mass m_e as a varying parameter we are implicitly considering a variation of m_e/Λ_{QCD}. In the last part of this thesis we will work with theories of grand unification. There, the grand unified scale M_{GUT} enters as natural scale which we choose to be constant. The appropriate conversion from a constant Λ_{QCD} to a constant M_{GUT} system of units is explained in Sec. 4.6.3.

2.5.1 The chiral limit

Many studies on varying constants work with the chiral limit, *i.e.* they assume that all quarks are massless [Epelbaum02, BeaneSavage02, Donoghue06]. Then all dimensionful QCD parameters are simply proportional to a power of Λ_{QCD}, which ameliorates their treatment considerably. For instance, QCD masses like the proton mass simply scale like

$$\Delta \ln m_p = \Delta \ln \Lambda_{QCD} \qquad (2.1)$$

and any other dimensionful QCD parameter according to its mass scale (for instance, cross sections with $[\sigma] = [\text{Energy}]^{-2}$ scale like $\Delta \ln \sigma = -2\Delta \ln \Lambda_{QCD}$). Switching on the quark masses, one obtains a finite range for pion-mediated interactions, which may greatly affect static and dynamical properties of nuclei. Also, the masses of all hadrons are affected at some order in chiral perturbation theory [Gasser82]. In this thesis we will work with the full quark contributions, which are for most QCD parameters known at least in first order chiral perturbation theory, *i.e.* to terms linear in the quark masses.

2.6 Probes of varying constants

A possible variation of constants can be tested in various ways. Common tests are laboratory based measurements, for instance of atomic transitions. Also, a multitude of astrophysical and cosmological effects can be studied under the question of constancy, which allow to probe physics over a timescale unreachable with laboratory measurements. In recent years probes of variations in the constants of nature have been performed with increasingly high accuracy. Whilst direct laboratory measurements do not point towards any variation, some astrophysical tests yield slight variations. In part III (Sec. 10) we will list all recent relevant probes of varying constants, followed by detailed studies on how one can combine the different outcomes in unified scenarios. BBN as a probe of varying constants will be examined in detail in part II of this thesis.

2.7 Fine-tuning of constants and the anthropic principle

Connected to the question of constancy of fundamental constants is the question of fine-tuning of these constants. Even though this question can be seen as a rather philosophical one, we will shortly comment on it.

As far as we know today, the value of most of the 27 fundamental constants is extremely fine-tuned in order to allow life to appear. It has been argued [Tegmark97] that even small deviations (less than or order of 1%) will make the appearance of any life impossible. For example, if the strong force was slightly weaker, multi-proton nuclei would not be stable, and if it was slightly stronger, hydrogen could fuse into helium-2. Similar arguments can be found for the electromagnetic and weak force and for many other natural constants.

This fine-tuning problem can be ameliorated, like all problems of this kind, by evoking the anthropic principle[7]. In short, this principle states that the Universe which we observe has to be capable to develop intelligent life like us. Otherwise we would not be here and could not ask the question why the Universe has exactly the laws of nature which it has. The final outcome is that the question why we are living in such a highly fine-tuned, $i.e.$ extremely improbable, universe has simply disappeared, because the actual probability we have to discuss is rather the probability under the condition of our existence, which is no longer vanishingly small.

In recent years scientists have come up with the idea of "multiverses", stating that universes with many different kinds of physical properties are constantly formed [Linde86]. This is supported by candidate theories of everything (like string theory) which ab initio do not seem to have hard constraints which would exclusively select our physics. Rather, they allow an extremely high number of different physical configurations. In the framework of those theories, universes with many different physical configurations bubble out constantly, and the anthropic principle states that our universe is the one of these many universes which allowed us to appear.

These considerations are not directly connected to the investigations which are subject of this thesis, except the fact that varying constants would lead to an even more fine-tuned universe: Not only the values of the constants today, but also their whole time evolution would need to be tuned such that we could appear. We will not comment on the point of fine-tuning in the following, but it has become clear that the problems we are tackling have some deeper connection to philosophy and the question of why we are actually here.

---

[7]The concept of the anthropic principle was systematically introduced by Brandon Carter in a contribution to a symposium honoring Copernicus' 500th birthday in 1973 [Carter74], even though the idea of the anthropic principle has already been used long before.

Chapter 3

Cosmology

In this thesis we will consider probes for varying constants from today back to the first minute after the Big Bang. Hence it is essential to understand the evolution of our Universe from the Big Bang until today. This chapter gives a short review of our current picture of the Universe, its history and present status, and the important equations that govern its evolution.

3.1 General relativity and the basics of cosmology

3.1.1 General relativity

General relativity is an extension of the theory of special relativity, which states that gravity is a purely geometric effect, generated by the curvature of spacetime. The relation between curvature and stress-energy is given by the Einstein field equations

$$R_{\mu\nu} - \frac{R}{2}g_{\mu\nu} = \frac{8\pi G_{\mathrm{N}}}{c^4}T_{\mu\nu}\,, \tag{3.1}$$

where $R_{\mu\nu}$ is the Ricci tensor, R the Ricci scalar, $g_{\mu\nu}$ the metric tensor and $T_{\mu\nu}$ the stress-energy tensor. Equation (3.1) is a complicated differential equation which can in general only be solved if one makes simplifying assumptions and/or uses numeric techniques.

3.1.2 The basics of cosmology

In cosmology one is interested in the evolution of the Universe as a whole. Thus, one usually confines oneself to physics on large scales which allows to make some simplifying assumptions that dramatically reduce the complexity of equation (3.1).

Assumption 1. *The main assumption of cosmology is that the Universe is homogeneous and isotropic on large scales.*

Of course, the existence of objects like the earth, sun *etc.* contradicts this assumption locally. However, if one averages over distances (> 1000 Mpc), it turns out that Assumption 1 is observationally well-justified[1]. Demanding all quantities to be

---
[1] The biggest known structure is the Sloan great wall which is 1.37 billion lightyears long.

Composition	w
non-relativistic matter	0
ultra-relativistic matter (radiation)	1/3
curvature	-1/3
cosmological constant	-1

Table 3.1: Equation-of-state parameters for different types of cosmological components

homogeneous and isotropic, one can show [WeinbergGRT] that the metric takes the simple form[2]

$$ds^2 = dt^2 - a(t)^2 \left(\frac{dr^2}{1-kr^2} + r^2 d\Theta^2 + r^2 \sin^2\Theta d\phi^2 \right), \tag{3.2}$$

where k describes the curvature and $a(t)$ is the scale parameter, related to the redshift z via

$$a = \frac{1}{1+z}. \tag{3.3}$$

The metric (3.2) is called Friedmann-Robertson-Walker metric (FRW metric). The scale parameter fulfills the Friedmann equations

$$H^2 := \left(\frac{\dot{a}}{a}\right)^2 = \frac{8\pi G_N}{3}\rho \tag{3.4}$$

$$3\frac{\ddot{a}}{a} = -4\pi G_N \left(\rho + \frac{3p}{c^2}\right), \tag{3.5}$$

where H is the Hubble constant and ρ and p denote the total energy and pressure density. These two densities are usually split up into the different components which are assumed to be present in today's Universe, baryonic and dark matter, dark energy (denoted with the symbol Λ), photons, neutrinos and curvature[3],

$$\rho = \rho_B + \rho_{DM} + \rho_\Lambda + \rho_\gamma + \rho_\nu + \rho_K. \tag{3.6}$$

The pressure is related to the energy via an equation of state,

$$p_i = w_i \rho_i, \tag{3.7}$$

where the equation-of-state parameter w_i depends on the composition of the components as shown in Table 3.1.

With the critical density defined as

$$\rho_C := \frac{3H^2}{8\pi G_N}, \tag{3.8}$$

---

[2]There are theories claiming that the averaged Einstein tensor $G_{\mu\nu} = R_{\mu\nu} - \frac{R}{2}g_{\mu\nu}$ which enters in Eq. (3.1) is not equivalent to the Einstein tensor derived from an averaged metric as given in Eq. (3.2). Since the actual outcome of these considerations is still unclear, we will not consider those theories in this thesis. See [Buchert07] for a recent review.

[3]In the very early Universe, also electrons will make a substantial contribution to the expansion rate. This applies to the epoch of BBN and will be explained in more detail in chapter 6.

3.1. GENERAL RELATIVITY AND THE BASICS OF COSMOLOGY

all densities are usually given as fractional densities

$$\Omega_i := \frac{\rho_i}{\rho_C}. \tag{3.9}$$

Note that equation (3.4) yields

$$\Omega_B + \Omega_{DM} + \Omega_\Lambda + \Omega_\gamma + \Omega_\nu + \Omega_K \equiv 1 \tag{3.10}$$

at all times.

In the course of the evolution of the Universe, the energy densities scale like

$$\rho \propto a^{-3(1+w)}, \tag{3.11}$$

which means that the values of Ω_i do not stay constant over time since we have different w for different kinds of energy densities (Tab. 3.1). As baryons and dark matter follow the same equation of state, one can combine these to the matter energy density

$$\Omega_M := \Omega_{DM} + \Omega_B. \tag{3.12}$$

Given today's values Ω_i^0, one can combine Eqs. (3.4), (3.8), (3.9) and (3.11) and use Tab. 3.1 to derive the time evolution of the Hubble constant,

$$H^2(a) = H_0^2 \left[\Omega_\gamma^0 a^{-4} + \Omega_M^0 a^{-3} + \Omega_K^0 a^{-2} + \Omega_\Lambda^0 \right], \tag{3.13}$$

where we have neglected the neutrinos which have no substantial contribution to today's content of the Universe[4] (see Tab. 3.2). Eq. (3.13) shows that at early times ($a \ll 1$) non-relativistic and relativistic matter become dominant and any cosmological constant component irrelevant, whilst at late times ($a \gg 1$) Ω_Λ dominates. The flow of cosmological components in the ΛCDM concordance model (see Sec. 3.2) is depicted in Fig. 3.1, where the time evolution of the fractional components is given by

$$\Omega_i = \frac{\Omega_i^0 a^{-3(1+w)}}{\Omega_\gamma^0 a^{-4} + \Omega_M^0 a^{-3} + \Omega_K^0 a^{-2} + \Omega_\Lambda^0}. \tag{3.14}$$

As can be seen in Fig. 3.1, today's Universe ($z = 0$) is dominated by dark energy (Ω_Λ) but did undergo 2 transitions, from radiation dominated to matter dominated and from matter to dark energy dominated:

- In the early Universe, the expansion was almost completely due to relativistic particles $\Rightarrow$ radiation-dominated era.

- At $z \approx 5000$, about 70,000 years after the Big Bang, we have matter-radiation-equality and the Universe becomes matter dominated.

- At $z \approx 0.4$, about 4.3 Gyrs ago, the Universe becomes dominated by dark energy (in a ΛCDM model).

---

[4]Further note that due to the tiny but non-vanishing mass of the neutrinos, the neutrino equation of state might change during the evolution of the universe.

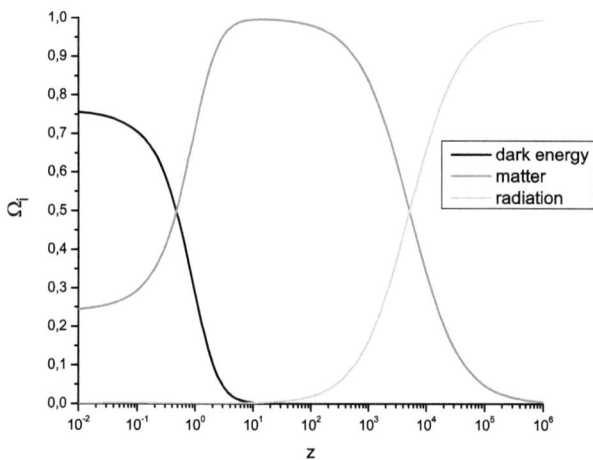

Figure 3.1: Evolution of the components in a ΛCDM Universe

3.2 The concordance model: Our current picture of the Universe

3.2.1 Historical development

Presumably, the question of where we come from and where we will go is as old as mankind. As a first modern physical approach to questions of origin, evolution and fate of the Universe, one usually considers Einstein's paper "Cosmological Considerations in the General Theory of Relativity" from 1917 [Einstein17]. One might say that high-precision observational cosmology started with the Hubble space mission in 1990. It was followed by further astrophysical and cosmological investigations, and basically all of these (mainly observational) tests point towards a coherent picture of our Universe, which is called the *"concordance model"*.

3.2.2 Our current picture of the Universe

According to the concordance model, the Universe started in a Big Bang[5] and has been expanding since then. All observational evidence points towards a so-called ΛCDM cosmology, stating that the Universe is geometrically flat ($\Omega_K \equiv 0$) and consists besides known baryonic matter, leptons and photons of an unknown "dark matter" component which has the property of non-relativistic, only gravitationally interacting heavy particles, and a "dark energy" component, in the simplest version described by

---

[5]Even though it is hoped that physics will once be able to explain the actual origin of this singular event, one is lacking an accepted theory of quantum gravitation which would allow to go beyond the time of the Planck epoch.

3.2. THE CONCORDANCE MODEL

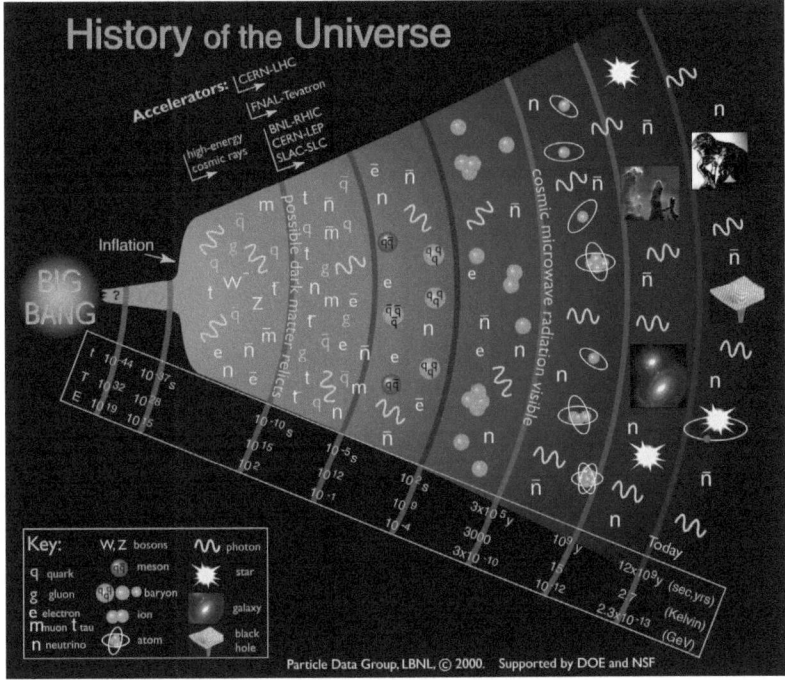

Figure 3.2: History of our Universe from Particle Data Group 2000

a cosmological constant Λ. See for instance [Bartelmann06] for a compilation of the major observational evidences for the Big Bang and [WMAP5] for recent parameter determinations including all major probes of the concordance model.

As can be seen in Fig. 3.1, we live in a dark-energy dominated universe just now but have undergone both radiation- and matter-dominated phases. The question why dark energy is taking over "just now" is unclear; this problem is called the "coincidence problem" or "why now problem" (see Sec. 5.1).

Looking back to the very beginning of our Universe, the concordance model suggests the following history of our Universe which is depicted in Fig. 3.2. Here, t is the age of the Universe, *i.e.* the time after the Big Bang, and T is the temperature of the Universe, defined by the photon temperature[6].

- $t \approx 10^{-43}$ seconds ($T \approx 10^{19}$ GeV):
  Planck epoch, needs to be described by a quantum theory of gravity.

- $t \approx 10^{-35}$ seconds ($T \approx 10^{15}$ GeV):
  Inflation (exponentially fast expansion of the Universe),
  Baryogenesis (production of matter-antimatter asymmetry)

---
[6]See Sec. 6.4.1 for details on the photon temperature.

- 10^{-6} seconds $\leq t \leq 10^{-2}$ seconds ($T \approx 0.1$ GeV):
  Quark-hadron transition: protons and neutrons form
- 1 second $\leq t \leq 3$ minutes, ($T \approx 1$ MeV, $z \approx 10^{10}$):
  Nucleosynthesis: light elements (D, He, Li) form
- $t \approx 70,000$ years ($T \approx 1$ eV, $z \approx 5000$):
  Beginning of matter dominated era
- $t \approx 300,000$ years ($T \approx 0.25$ eV, $z \approx 1100$):
  Recombination, the cosmic microwave background (CMB) forms
- After that:
  Galaxy/star formation.

Experimental tests of high-energy physics currently do not go beyond the TeV region. The processes which happened during the Planck epoch, inflation and also baryogenesis go beyond the physics of the Standard Model and are hence subject to a lot of speculation. Also the quark-hadron transition is hard to describe since it involves QCD at low energies, in the regime where QCD effects cannot be evaluated perturbatively. Hence, the oldest cosmological events which can be reasonably described quantitatively with known physics are primordial nucleosynthesis and CMB formation.

3.3 Cosmological parameter values

In recent years, important and quite extensive missions have been undertaken to deepen our understanding of cosmological relations. In particular, WMAP, SDSS and Supernovae Ia [WMAP5, HinshawWMAP5](see also references therein) have yielded a coherent set of cosmological parameters of a precision which had been inconceivable 10 years ago. However, compared to the Standard Model of particle physics, the concordance model of cosmology is rather new and by far less tested. The set of cosmological parameters for a ΛCDM cosmology is given in Tab. 3.2. It turns out that only 4.6% of our Universe is made of "known" ordinary baryonic matter, the rest of the Universe is dark matter and dark energy. The energy composition of our Universe today is shown in Fig. 3.3.

Using the evolution equations of Sec. 3.1, the present-day values of cosmological parameters allow to deduce the content of our Universe in the past and also in the future[7]. As observations also allow us to look back in time, the picture for the past is nowadays quite clear and observationally probed. The composition history of our Universe in a ΛCDM model is shown in Fig. 3.1.

---

[7]Cosmological models allow to extrapolate cosmology into the future, however models are not tested sufficiently to allow a definite prediction of what the ultimate fate of the Universe will be.

3.3. COSMOLOGICAL PARAMETER VALUES

Quantity	Symbol	Value
Hubble expansion rate	H_0	$100\,h\,\mathrm{km/s\,Mpc^{-1}}$
normalized Hubble expansion rate	h	0.701 ± 0.013
baryon density	$\Omega_b \equiv \rho_b/\rho_c$	0.0462 ± 0.0015
dark matter density	$\Omega_{dm} \equiv \rho_{dm}/\rho_c$	0.233 ± 0.013
matter density	$\Omega_m \equiv \Omega_b + \Omega_{dm}$	0.279 ± 0.013
dark energy density	$\Omega_\Lambda \equiv \rho_\Lambda/\rho_c$	0.721 ± 0.015
radiation density	$\Omega_\gamma \equiv \rho_\gamma/\rho_c$	$(5.0 \pm 0.2) \cdot 10^{-5}$
neutrino density	$\Omega_\nu \equiv \rho_\nu/\rho_c$	$< 0.013\ (95\%\,\mathrm{CL})$
baryon to photon ratio	$\eta \equiv n_b/n_\gamma$	$(6.21 \pm 0.16) \cdot 10^{-10}$
CMB temperature	T	$2.725\,\mathrm{K}$

Table 3.2: Parameters describing our Universe. WMAP "recommended parameter values" [WMAP5, HinshawWMAP5] from WMAP5, BAO and SN for a ΛCDM cosmology

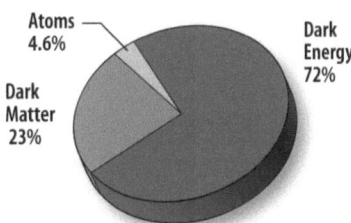

Figure 3.3: Today's energy content of our Universe. From NASA / WMAP Science Team

Chapter 4

The Standard Model and beyond

4.1 The Standard Model of particle physics

The Standard Model (SM) of particle physics describes the elementary particles and three of the four fundamental interactions, the strong, weak and electromagnetic interaction. This section will only give a rough overview over some particular aspects of the Standard Model which will be of relevance later. For a more comprehensive introduction, see for instance [HalzenMartin] or any other textbook on modern particle physics.

The list of the fundamental particles of the Standard Model comprises

- six leptons $(e, \mu, \tau, \nu_e, \nu_\mu, \nu_\tau)$,
- six quarks (u, d, s, c, b, t),
- the gauge bosons as mediators of the fundamental interactions,
- a Higgs boson.

The matter particles enter as pointlike massless fermions, and the interactions are introduced by demanding a local $SU(3) \times SU(2) \times U(1)$ gauge symmetry. Here, the group $SU(3)$ is responsible for quantum chromodynamics (QCD), with 8 massless gauge bosons called *gluons* as mediators of the strong force. At small momenta, the strong coupling constant becomes large (see Sec. 4.2), which is thought to be the explanation for *confinement*, *i.e.* the fact that only color-neutral particles are observed in nature.

The theory of electroweak interaction goes back to the seminal work of Glashow, Salam and Weinberg [Weinberg67, Salam69, Glashow70] (Nobel Prize 1979). It describes the electroweak interaction by a $SU(2) \times U(1)$ gauge symmetry which is broken by the Higgs mechanism into the weak interaction (with massive gauge bosons $W^\pm$ and Z^0) and the electromagnetic sector with the photon as massless mediator of the electromagnetic force. In particular, the $SU(2) \times U(1)$ gauge symmetry implies four massless gauge bosons, written as $W_\mu^\pm$, W_μ^3 and B_μ. Additionally, one introduces a scalar Higgs field ϕ (as a weak doublet under $SU(2)$ which has altogether 4 real

4.2. RUNNING OF COUPLINGS

components) and gives it a potential $V(\phi)$ which results in a vacuum that is not symmetric under the $SU(2) \times U(1)$ gauge symmetry. This leads to a spontaneous symmetry breakdown of the electroweak symmetry, and the Higgs field obtains a nonzero vacuum expectation value $\langle\phi\rangle \approx 246$ GeV. One of the four Higgs components becomes a massive scalar particle, which is the only particle of the SM which has not yet been observed. The $W_\mu^\pm$ and a combination of W_μ^3 and B_μ obtain masses proportional to $\langle\phi\rangle$ and become the massive mediators of the weak force, $W^\pm$ and Z^0, where the three remaining components of the Higgs form the longitudinal modes of the $W^\pm$ and Z^0. The coupling constant α_{em} of the remaining electromagnetic symmetry group $U(1)_{em}$ can be obtained from the coupling constants of the original $SU(2) \times U(1)$ coupling constants α_1, α_2 (see e.g. [HalzenMartin]),

$$\alpha_{em}^{-1} = \alpha_1^{-1} + \alpha_2^{-1} . \tag{4.1}$$

Also, the SM fermions obtain masses via the Higgs mechanism, their mass is a product of Higgs v.e.v. and a Yukawa coupling h_i, for instance for the electron

$$m_e = h_e \langle\phi\rangle . \tag{4.2}$$

4.2 Running of couplings

The influence of fluctuations with different momenta leads to scale dependent coupling constants. See for instance [Wilson71, Wegner72, Wilson73] or any good textbook on quantum field theory for details of this process. Generally, physical systems at slightly different scales are described by the similar laws of physics, with slightly changed parameters. In quantum field theory, this behavior is described by the famous beta function, which describes the behavior of the coupling parameter g under slight changes of the energy scale μ,

$$\mu \frac{\partial g}{\partial \mu} = \beta(g) . \tag{4.3}$$

Using the coupling constant $\alpha := \frac{g^2}{4\pi}$ instead, one can also define a β function for α,

$$\mu \frac{\partial \alpha}{\partial \mu} = \beta(\alpha) . \tag{4.4}$$

The mathematical apparatus to investigate these changes of physical systems under scale transformations is called the renormalization group (RG). In quantum field theory, the renormalization group equation (4.3) can only be computed perturbatively as the exact RG equation would in principle include an infinite order of loop corrections. For our purpose the first-order (one-loop) RG equations are sufficient, and these are known for all three interactions of the Standard Model. In particular, the beta function for QED (with only photons and electrons present) at first order is given by

$$\beta_{em}(\alpha_{em}) = \frac{2\alpha_{em}^2}{3\pi} , \tag{4.5}$$

which is solved by

$$\alpha_{em}(\mu) = \frac{\alpha_{em}(\mu_0)}{1 - \frac{2\alpha_{em}(\mu_0)}{3\pi} \ln\left(\frac{\mu}{\mu_0}\right)} . \tag{4.6}$$

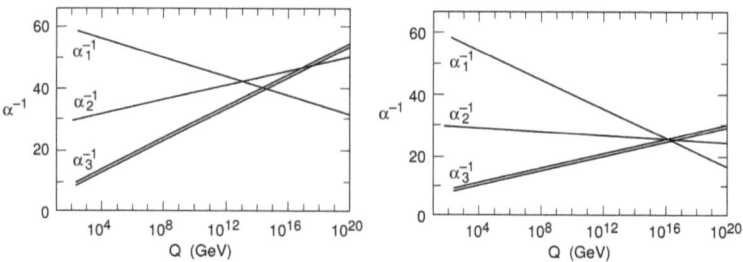

Figure 4.1: Running of coupling constants of the three gauge groups $SU(3) \times SU(2) \times U(1)$ in the Standard Model (left) and in SUSY (right). α_1 is scaled by a factor $\frac{5}{3}$, see Eq. (4.14). From [Peskin97].

The fine structure constant α is defined in the limit of zero momentum transfer, i.e. for $\mu \leq m_e$. For QCD, the beta function is

$$\beta_{strong}(\alpha_S) = -\left(11 - \frac{2n_f}{3}\right)\frac{\alpha_S^2}{2\pi}, \qquad (4.7)$$

solved by

$$\alpha_S(\mu) = \frac{\alpha_S(\mu_0)}{1 + \frac{\alpha_S(\mu_0)}{6\pi}(33 - 2n_f)\ln\left(\frac{\mu}{\mu_0}\right)}. \qquad (4.8)$$

Here n_f is the number of quark flavors present, i.e. the number of quarks with mass $m_q \leq \mu$. As $n_f \leq 6$ in the Standard Model, the beta function β_{strong} is negative.

The running of coupling constants is shown in Fig. 4.1. As the beta function for QCD is negative, the QCD coupling diverges when going to low energies. This effect, which was found by Wilczek, Politzer and Gross (Nobel price 2004), is thought to be the reason for confinement. As opposed to the electroweak theory, QCD thus has an intrinsic energy scale induced by the RG equation, the scale where α_S becomes formally infinite. Choosing $m_s < \mu_0 < m_c$ such that n_f remains constant ($n_f \equiv 3$), this happens when the denominator in equation (4.8) becomes zero, i.e.

$$\frac{\alpha_S(\mu_0)}{6\pi}(33 - 2n_f)\ln\left(\frac{\mu}{\mu_0}\right) = -1, \qquad (4.9)$$

which happens at the *QCD invariant scale* $\mu \equiv \Lambda_{QCD}$, defined by

$$\Lambda_{QCD} := \mu_0 \exp\left(\frac{-6\pi}{(33 - 2n_f)\alpha_S(\mu_0)}\right). \qquad (4.10)$$

Hence Eq. (4.8) can be rewritten as ($\mu < m_c$)

$$\alpha_S(\mu) = \frac{6\pi}{(33 - 2n_f)\ln(\mu/\Lambda_{QCD})}. \qquad (4.11)$$

Note that when the energy μ becomes of the order of Λ_{QCD}, perturbation theory breaks down, and a world of quarks and gluons becomes a world of pions, protons

4.3. THE NECESSITY OF A "THEORY BEYOND"

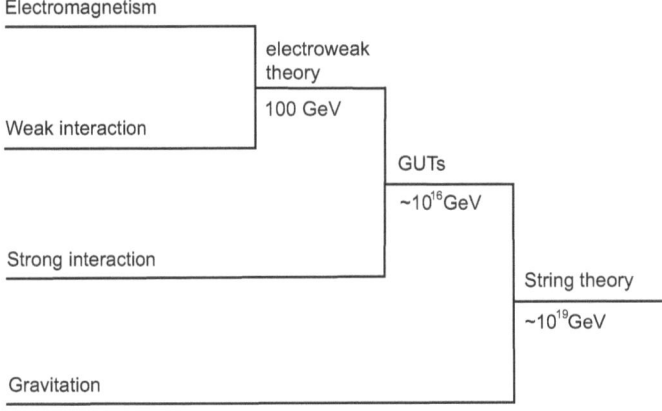

Figure 4.2: Unification of the four forces in the string theory picture

and so on. This is revealed in the fact that Λ_{QCD} is of the order of light meson masses [Berger06],

$$\Lambda_{QCD} \approx 200 \text{ MeV} . \tag{4.12}$$

The beta functions given in Eqs. (4.5) and (4.7) are simplified versions. In the first case, Eq. (4.5) only holds for one particle present, and in the second case one has to note that n_f is not constant at all energies μ. In the full functions, particles only contribute when energies are above the particle's threshold energies, which are typically the corresponding particle masses. Hence every particle contributes one threshold term to the renormalization group equation. We will give the full renormalization group equations, including extra terms coming from additional supersymmetric particles (see Sec. 4.4) in Sec. 4.5.

4.3 The necessity of a "theory beyond"

From the point of a theorist, the established Standard Model of particle physics cannot be the end of the story. One can definitely say that at latest at the Planck energy scale $M_P \approx 10^{19}$ GeV quantum gravity effects will become important, demanding a quantized description of gravity. A further hint that the Standard Model of particle physics might not be the end of the story is the running of the coupling constants. They seem to meet at a energy scale of $M_{GUT} \approx 10^{16}$ GeV as depicted in Fig. 4.1. Hence, it appears likely that electroweak and strong interaction can be unified in a grand unified theory (GUT). Within such grand unified theories, it is most likely that if any coupling constant of the Standard Model varies, all coupling constants will vary. In the later chapters, we will assume that some kind of GUT is realized, and hence the electroweak and strong coupling constants are related to each other. Further details on grand unified theories are given in Sec. 4.5.

Actually, one can even go further with the idea of unification. String theory, for instance, implements an unification of the interactions of the Standard Model and also gravity. The unification of the four forces including gravity in a string theory picture is shown in Fig. 4.2. In such theories, parameters related to the mass of the SM particles (*e.g.* the Yukawa couplings) should also derive from some sort of unified greater theory, whereas at the level of the Standard Model and also at the level of simple GUTs, there is no direct relation to the gauge coupling sector[1]. In this thesis, we will assume constant Yukawa couplings.

4.4 Supersymmetry and the MSSM

Many high-energy theories (*e.g.* string theory) contain supersymmetry as an essential part of the theory. Supersymmetry establishes a symmetry between bosons and fermions. Every boson gets a fermionic partner and vice versa with the same quantum numbers. For this thesis the motivation and theoretical framework of supersymmetry are not needed, so I will refrain from going into too much detail. Introductions to supersymmetry can be found in many textbooks and reviews, *e.g.* [NillesSUSY].

Within the Standard Model of particle physics, no supersymmetric partner can be found, hence in a supersymmetric extension of the Standard Model one has to introduce additional supersymmetric partners for every single particle of the SM. These supersymmetric partners are assumed to be heavier than the current experimentally tested mass region (≈ 100 GeV).

The minimal supersymmetric extension of the Standard Model is called the *MSSM (minimal supersymmetric standard model)*. It contains an additional supersymmetric partner for every SM particle. Furthermore, with supersymmetry, a single Higgs doublet would result in a gauge anomaly, so a second Higgs doublet is introduced. Hence the MSSM contains 2 additional (heavy) neutral Higgs scalars and two charged Higgs scalars, supplemented by the appropriate superpartners.

The complete particle spectrum of the MSSM is given in Tab. 4.1. When working with supersymmetric theories in this thesis, we will always assume that the MSSM particle spectrum is realized.

4.5 Grand unification

In grand unified theories, the gauge group of the Standard Model $SU(3) \times SU(2) \times U(1)$ with coupling constants g_S, g_2, g_1 is unified into a bigger Lie group (*e.g.* $SU(5)$ or $SO(10)$) with a single coupling constant g_X at a certain energy scale,

$$M_{GUT} \approx 10^{16} \text{ GeV} ,\qquad(4.13)$$

which is assumed an independent parameter and can also vary with time. Its actual value depends on the specific form of the grand unified theory (*e.g.* SUSY/non-SUSY).

---

[1] Due to renormalization group effects, also the Yukawa couplings get contributions from coupling parameters. However, we will show in Sec. 11.1 that these effects are small and can be neglected when studying variations of parameters.

4.5. GRAND UNIFICATION

SM particle	El. charge	Spin	SUSY partner	Spin
quarks u,c,t	2/3	1/2	squarks $\tilde{u}, \tilde{c}, \tilde{t}$	0
quarks d,s,b	-1/3	1/2	squarks $\tilde{d}, \tilde{s}, \tilde{b}$	0
charged leptons e, μ, τ	-1	1/2	sleptons	0
neutrinos ν_e, ν_μ, ν_τ	0	1/2	sneutrinos ($\tilde{\nu}$)	0
photon γ	0	1	photino $\tilde{\gamma}$	1/2
Z^0	0	1	Zino ($\tilde{Z}$)	1/2
Z^0 neutral Higgs scalar	0	0	Zino Higgs	1/2
$W^\pm$	± 1	1	Wino ($\tilde{W}$)	1/2
$W^\pm$ charged Higgs scalar	± 1	0	Wino Higgs	1/2
8 gluons	0	1	8 gluinos ($\tilde{g}$)	1/2
neutral Higgs H^0	0	0	higgsino ($\tilde{H}^0$)	1/2
2 MSSM neutral Higgs	0	0	2 neutral higgsinos	1/2
2 MSSM charged Higgs $H^\pm$	± 1	0	2 charged higgsinos $\tilde{H}^\pm$	1/2

Table 4.1: The MSSM particle spectrum

It can be shown (see *e.g.* [WeinbergQFT2]) that for any unification of $SU(3) \times SU(2) \times U(1)$ with couplings $g_3 \equiv g_S$, g_2 and g_1 into a simple Lie group with coupling g_X, one obtains the relation

$$g_X^2 = g_S^2 = g_2^2 = \frac{5}{3}g_1^2 \qquad (4.14)$$

which holds at $E = M_{GUT}$. For the electromagnetic interaction, Eqs. (4.1) and (4.14) yield

$$\alpha_{em}^{-1} = \alpha_2^{-1} + \alpha_1^{-1} = \alpha_X^{-1} + \frac{5}{3}\alpha_X^{-1} = \frac{8}{3}\alpha_X^{-1} , \qquad (4.15)$$

which actually only holds at $E = M_{GUT}$ but will be of relevance when studying the running of α_{em} in a GUT framework.

The value of the unified coupling α_X can roughly be estimated from the RG running as displayed in Fig. 4.1, showing that α_X^{-1} is of the order $38 - 45$ in the non-SUSY case and $23 - 29$ in the SUSY case [Amaldi91]. We take as representative values [Dent03]

$$\alpha_X = 1/40 \quad \text{(non-SUSY)} \qquad (4.16)$$
$$\alpha_X = 1/24 \quad \text{(SUSY)}. \qquad (4.17)$$

At lower energies, the GUT symmetry is broken and the relation (4.14) does not hold any longer. The coupling constants of the SM evolve separately according to the renormalization group equations (see section 4.2). Generalizing the running of couplings as given in equations (4.6) and (4.8) to the full SM/MSSM particle spectrum, we obtain for QCD

$$\alpha_S^{-1}(\mu) = \alpha_X^{-1} - \frac{1}{2\pi}\sum_i b_i \ln \frac{\mu}{M_{GUT}} - \frac{1}{2\pi}\sum_{th} b^{th} \ln\left(\frac{m^{th}}{M_{GUT}}\right) \qquad (4.18)$$

where the first sum goes over all particles i with threshold mass $m^{th} < \mu$ and the second sum goes over all particles with $m^{th} > \mu$. For the b_i and b^{th} the values from Tab. 4.2 have to be applied.

Type of particle	$b^{(th)}$
quarks	2/3
gluons	-11/8
squarks	1/3
gluinos	1/4

Table 4.2: Renormalization group coefficients for the strong interaction

Type of particle	f^{th}
chiral (or Majorana) fermion	2/3
complex scalar	1/3
vector boson	-11/3

Table 4.3: Renormalization group coefficients for the electromagnetic interaction

The corresponding expression for the fine-structure constant $\alpha := \alpha_{em}(m_e)$ is

$$\alpha^{-1} = \frac{8}{3}\alpha_X^{-1} - \frac{1}{2\pi} \sum_{th} Q_{em}^{th^2} f^{th} \ln\left(\frac{m^{th}}{M_{GUT}}\right), \qquad (4.19)$$

where the factor $\frac{8}{3}$ derives from Eq. (4.15), Q_{em} denotes the electric charge and for f^{th} the values given in Tab. 4.3 are applied. An analogous equation also holds for the weak coupling, where the electric charge is replaced by the weak isospin[2]. When dealing with weak interactions in this thesis, we will only be working with terms that contain the weak coupling in terms of the Fermi constant[3],

$$G_F = \frac{\sqrt{2}}{8} \frac{g_w^2}{M_W^2}. \qquad (4.20)$$

As $M_W = \frac{g_w \langle \phi \rangle}{2}$ with $\langle \phi \rangle$ the Higgs v.e.v., the weak coupling g_w drops out and the Fermi constant can be expressed only in terms of the Higgs v.e.v.,

$$G_F = \frac{1}{\sqrt{2}\langle \phi \rangle^2}. \qquad (4.21)$$

4.6 Variations in a GUT framework

The GUT relations which were introduced in the preceding section show that within a GUT framework, the coupling constants are usually related to further fundamental parameters, in particular the GUT coupling α_X and threshold masses. In this section we will derive the equations which relate variations in the GUT coupling constant α_X and particle masses to variations in the SM coupling constants.

---

[2] For the SU(2) weak interaction, the weak isospin effectively acts like a multiplicative charge factor, hence it can be treated analogously to the electric charge.
[3] See for instance the weak decay of the neutron, Sec. 8.1.3.

4.6.1 Variation of the electromagnetic coupling

For the MSSM particle spectrum, we obtain for the fine-structure constant from Eq. (4.19)

$$\alpha^{-1} = \frac{8}{3}\alpha_X^{-1} - \frac{1}{2\pi}\left[\frac{4}{3}\cdot 3\cdot\left(\left(\frac{2}{3}\right)^2 + 2\left(\frac{1}{3}\right)^2\right)\ln\frac{\Lambda_{QCD}}{M_{GUT}} + \frac{4}{3}\cdot 3\left(\frac{2}{3}\right)^2\ln\frac{m_c m_t}{M_{GUT}^2}\right.$$

$$+\frac{4}{3}\cdot 3\left(\frac{1}{3}\right)^2\ln\frac{m_b}{M_{GUT}} + \frac{4}{3}(1)^2\ln\frac{m_e m_\mu m_\tau}{M_{GUT}^3} + \left(-\frac{11}{3}\cdot 2 + \frac{1}{3}\right)\ln\frac{M_W}{M_{GUT}}$$

$$+\frac{2}{3}\cdot 3\cdot(1)^2\ln\frac{m_{\tilde{l}}}{M_{GUT}} + \frac{2}{3}\left(3\cdot 3\cdot\left(\frac{2}{3}\right)^2 + 3\cdot 3\cdot\left(\frac{1}{3}\right)^2\right)\ln\frac{m_{\tilde{q}}}{M_{GUT}}$$

$$\left. +\frac{2}{3}\cdot 2\cdot(1)^2\ln\frac{m_{\tilde{W}}}{M_{GUT}} + \frac{2}{3}\cdot 2\cdot(1)^2\ln\frac{m_{\tilde{H}}}{M_{GUT}} + \frac{1}{3}\ln\frac{m_{H^\pm}}{M_{GUT}}\right] \quad (4.22)$$

where it has been used that

- The light quarks u, d, s decouple at $m^{th} = \Lambda_{QCD}$.
- The quarks enter in 3 different colors
- The charged leptons enter as both left- and righthanded particles (2 chiral fermions)
- The massive gauge bosons $W^\pm$ have to be supplemented by a charged complex Higgs scalar (longitudinal DOFs)
- $m_{H^\pm}$ is the mass of the additional charged Higgs scalars which have to be introduced in MSSM.

Taking the linear variation of Eq. (4.22), we obtain for the variation of the fine structure constant, including the MSSM particles,

$$\frac{\Delta\ln\alpha}{\alpha} = +\frac{8}{3}\frac{\Delta\ln\alpha_X}{\alpha_X} + \frac{1}{2\pi}\left(\frac{8}{3}\Delta\ln\frac{\Lambda_{QCD}}{M_{GUT}} + \frac{16}{9}\Delta\ln\frac{m_c m_t}{M_{GUT}^2} + \frac{4}{9}\Delta\ln\frac{m_b}{M_{GUT}}\right.$$

$$\left. +\frac{4}{3}\Delta\ln\frac{m_e m_\mu m_\tau}{M_{GUT}^3} - \frac{21}{3}\Delta\ln\frac{M_W}{M_{GUT}} + 8\Delta\ln\frac{\tilde{m}}{M_{GUT}} + \frac{1}{3}\Delta\ln\frac{m_{H^\pm}}{M_{GUT}}\right). \quad (4.23)$$

Not knowing the actual mass or mass generating mechanism for the superpartners, we assume that the mechanism is the same for all superpartners and define $\tilde{m}$ as the average superpartner mass. To obtain the corresponding relation in nonsupersymmetric models, one simply has to leave out the terms with $\tilde{m}$ and $m_{H^\pm}$.[4] When later dealing with variations in supersymmetric models, we will further assume $\Delta\ln\tilde{m} = \Delta\ln m_{H^\pm}$, so the last two terms can be combined into one.

---

[4]Note that the RG equations (4.22), (4.23) and also (4.24) and (4.26) only hold under the condition that all threshold masses are smaller than M_{GUT}. Hence, the nonsupersymmetric case is *not* obtained in the limit $m_{susy} \to \infty$.

4.6.2 Variation of the QCD scale

As α_S diverges at $\mu = \Lambda_{QCD} \approx 200$ MeV, we are interested in the value of $\alpha_S(\mu)$ in the regime $\Lambda_{QCD} < \mu < m_c$ (in this regime, $n_f = 3$). For the MSSM particle spectrum, we obtain

$$\alpha_S^{-1}(\mu) = \alpha_X^{-1} + \frac{9}{2\pi}\ln\left(\frac{\mu}{M_{GUT}}\right) - \frac{1}{2\pi}\left(\frac{2}{3}\ln\frac{m_c m_b m_t}{M_{GUT}^3} + \frac{6}{3}\ln\frac{m_{\tilde{q}}}{M_{GUT}} + \frac{8}{4}\ln\frac{m_{\tilde{g}}}{M_{GUT}}\right). \tag{4.24}$$

Inserting Eq. (4.24) into Eq. (4.10) ($\Lambda_{QCD} < \mu_0 = \mu < m_c$, $n_f = 3$) yields

$$\frac{\Lambda_{QCD}}{M_{GUT}} = e^{-2\pi/9\alpha_X}\left(\frac{m_c m_b m_t}{M_{GUT}^3}\right)^{2/27}\left(\frac{m_{\tilde{q}} m_{\tilde{g}}}{M_{GUT}}\right)^{2/9} \tag{4.25}$$

and the linear variation gives

$$\Delta\ln\frac{\Lambda_{QCD}}{M_{GUT}} = \frac{2\pi}{9\alpha_X}\Delta\ln\alpha_X + \frac{2}{27}\Delta\ln\frac{m_c m_b m_t}{M_{GUT}^3} + \frac{2}{9}\left(\Delta\ln\frac{m_{\tilde{q}}}{M_{GUT}} + \Delta\ln\frac{m_{\tilde{g}}}{M_{GUT}}\right). \tag{4.26}$$

When later dealing with variations in supersymmetric models, we will further assume $\Delta\ln m_{\tilde{q}} = \Delta\ln m_{\tilde{g}}$, so the last two terms can be combined into one.

4.6.3 Conversion of units

As has been explained in Sec. 2.5, we will work with two different systems of units. During the discussion of BBN processes, we choose units with $\Lambda_{QCD} = const.$ as the BBN energy scale is of roughly the same order of magnitude. When applying grand unified theories, M_{GUT} is the more appropriate energy scale to keep constant. However, we usually neglect the reference scale and write $\Delta\ln m_e$ instead for $\Delta\ln\frac{m_e}{\Lambda_{QCD}}$, for instance. The conversion to a different system is then performed by keeping track of all reference scales,

$$\Delta\ln\frac{m_e}{M_{GUT}} = \Delta\ln\frac{m_e}{\Lambda_{QCD}} + \Delta\ln\frac{\Lambda_{QCD}}{M_{GUT}}. \tag{4.27}$$

Obviously, the term $\Delta\ln\frac{\Lambda_{QCD}}{M_{GUT}}$ naturally enters when converting from the $\Lambda_{QCD} = const.$ to the $M_{GUT} = const.$ system of units, its explicit dependence on the unified coupling and particle masses is given in Eq. (4.26). Keep in mind, however, that the reference scale has to enter in the correct power, for instance the gravitational constant has units $[G_N] = [\text{Energy}]^{-2}$, hence it enters as $\Delta\ln G_N \Lambda_{QCD}^2$.

Chapter 5

Models of quintessence

5.1 Problems of the cosmological constant

Recent observations show that roughly 75% of the energy content of our Universe is made from dark energy (see Sec. 3.2). However, the nature of dark energy is still far from being clear. The assumption that the cosmological constant derives from a vacuum energy density suffers from a severe fine-tuning problem. In particular, the oberseved dark energy density evaluates to [Copeland06]

$$\rho_\Lambda = \frac{\Lambda M_P^2}{8\pi} \approx 10^{-47} \text{ GeV}^4, \qquad (5.1)$$

while the vacuum energy density of particle physics which is evaluated by summing up the zero-point energies of the present quantum fields gives [Copeland06]

$$\rho_{vac} \approx \frac{M_P^4}{16\pi^2} \approx 10^{74} \text{ GeV}^4. \qquad (5.2)$$

Here we have chosen M_P as a natural cut-off scale where we assume that the known quantum field theory is no longer applicable. Obviously, there is a discrepancy of the order of 10^{121}. Assuming that the dark energy comes from a particle physics origin, one would have to introduce counter terms which have to be extremely fine-tuned. Hence this problem is called the "finetuning" problem.

A further problem which is related with dark energy can be seen in Fig. 3.1. In a ΛCDM model, the dark energy is only recently becoming important, and the time when the universe switched from a matter-dominated to a dark energy dominated epoch is only 4.3 billion years ago. There is no natural reason why these two presumably completely independent constituents of our Universe are of about the same order of magnitude and / or why we live in a period of time where this is the case. This problem is called the "coincidence problem" or "why now problem", and typically cosmological models with a cosmological constant (like the ΛCDM model) fail to address this issue.

5.2 Basics of quintessence

Models of quintessence [Wetterich88.1, RatraPeebles88] can offer an explanation to the issues mentioned in the previous section. A good review on dark energy models can be found in [Copeland06].

CHAPTER 5. MODELS OF QUINTESSENCE

In quintessence theories, one introduces a scalar field φ (called the cosmon) which is coupled to gravity and, most times, also to matter and gauge fields. A typical Lagrangian for a quintessence theory including couplings to matter and the electromagnetic gauge field looks like [Wetterich02.1, Copeland06]

$$\mathcal{L} = \bar{M}_{\mathrm{P}}^2 R + \frac{1}{2}(\partial\varphi)^2 + V(\varphi) - V_{\varphi m} + \mathcal{L}_{\mathrm{em}} \tag{5.3}$$

with gauge field coupling

$$\mathcal{L}_{\mathrm{em}} = -\frac{1}{4}(1 + \lambda_{\mathrm{em}}\varphi)^{-1} F_{\mu\nu} F^{\mu\nu} \tag{5.4}$$

and matter term

$$V_{\varphi m} = m_e(\varphi)\bar{e}e + m_u(\varphi)\bar{u}u + ... + m_{\mathrm{dark}}(\varphi)\overline{\Psi_{\mathrm{dark}}}\Psi_{\mathrm{dark}} + ... \tag{5.5}$$

Here $\bar{M}_{\mathrm{P}}$ denotes the reduced Planck mass,

$$\bar{M}_{\mathrm{P}} = \frac{1}{\sqrt{8\pi}} M_{\mathrm{P}} = \frac{1}{\sqrt{8\pi G_{\mathrm{N}}}} \,. \tag{5.6}$$

If there are only slight changes in the cosmon field, the dependence of the mass, $m(\varphi)$, can be linearized, *i.e.*

$$m(\varphi) = (1 + \lambda\varphi)m_0 \tag{5.7}$$

with some coupling λ. However, there are also models with significant changes in the masses, for instance in the models of growing neutrinos in Sec. 5.4, where we apply a more advanced nonlinear expression.

In order to derive one of the main properties of quintessence, its capability of producing accelerated expansion, we can neglect couplings to matter and gauge fields and work instead with the action

$$S = \int d^4x \sqrt{-g} \left[-\frac{1}{2}(\partial\varphi)^2 - V(\varphi) \right] . \tag{5.8}$$

In the background of a flat FRW cosmology (Sec. 3.1.2), and assuming that φ is homogeneous, *i.e.* it only depends on time, a variation of the action (5.8) with respect to φ yields the equation of motion

$$\ddot{\varphi} + 3H\dot{\varphi} + \frac{\mathrm{d}V}{\mathrm{d}\varphi} = 0 \,. \tag{5.9}$$

The corresponding energy momentum tensor

$$T_{\mu\nu} = \frac{-2}{\sqrt{-g}} \frac{\delta S}{\delta g^{\mu\nu}} \tag{5.10}$$

yields the energy and pressure densities

$$\rho = -T_0^0 = \frac{1}{2}\dot{\varphi}^2 + V(\varphi) \tag{5.11}$$

$$p = T_i^i = \frac{1}{2}\dot{\varphi}^2 - V(\varphi) \,. \tag{5.12}$$

Then Eqs. (3.4) and (3.5) yield the relations

$$H^2 = \frac{8\pi G_N}{3}\left[\frac{1}{2}\dot{\varphi}^2 + V(\varphi)\right] \qquad (5.13)$$

$$\frac{\ddot{a}}{a} = -\frac{8\pi G_N}{3}\left[\dot{\varphi}^2 - V(\varphi)\right], \qquad (5.14)$$

showing that we get an accelerating universe when $\dot{\varphi}^2 < V(\varphi)$. Introducing the kinetic energy $T := \dot{\varphi}^2/2$, we define the equation of state parameter of quintessence,

$$w_h := \frac{p}{\rho} = \frac{T-V}{T+V}. \qquad (5.15)$$

In the next section we introduce two specific models of quintessence, crossover quintessence which has been introduced 20 years ago [Wetterich88.1] and a very recent model, where quintessence is strongly coupled to neutrinos.

As the cosmon also couples to other fields and matter, one question one might ask is whether the cosmon evolution decouples in a local cluster with high "cosmon charge density" from the cosmological evolution. It has been shown [Wetterich02.1, Mota03, Shaw05] that for a very light field weakly coupled to matter the local perturbations are generally small relative to the cosmological evolution. In other words, the evolution of the scalar field in a cluster of galaxies or on Earth does not decouple from the cosmological evolution (in distinction to the gravitational field), such that its cosmological time evolution is reflected in a universal variation of couplings, both on Earth and in the distant Universe.

5.3 Crossover quintessence models

Our first class of models is "crossover quintessence" [Hebecker00, Doran07, Wetterich02.2]. Here the scalar field follows tracking solutions [Wetterich88.1, RatraPeebles88] at large redshift. In this early epoch the equation of state w_h is equal to that of the dominant energy component (matter or radiation). One particular difference to cosmological constant models is that this type of quintessence models yields a non vanishing amount of early dark energy, $\Omega_{h,e}$ [1]. Typically, such models have an exponential potential, for instance

$$V(\varphi) = \bar{M}_P^4 e^{-\alpha \frac{\varphi}{M_P}}. \qquad (5.16)$$

In this specific case, α is related to the early dark energy fraction [Wetterich88.1, Amendola08],

$$\Omega_{h,e} = \frac{n}{\alpha^2} \qquad (5.17)$$

with $n = 3(4)$ for the matter (radiation) epoch. Late-time acceleration can be achieved, for instance, by slight modifications in the cosmon potential or, equivalently, in the kinetic term [Hebecker00].

At some intermediate redshift before the onset of acceleration, the time evolution of the cosmon slows down. In consequence, there is a crossover to a negative equation of state and the fraction of energy density due to the scalar begins to grow. In recent epochs the field has an effective equation of state $w_h \gtrsim -1$. The aim of this thesis is

---

[1] The effects of early dark energy on the measurements of baryon acoustic peaks have been studied by the author in [DST06]. However, these considerations are not subject of this thesis.

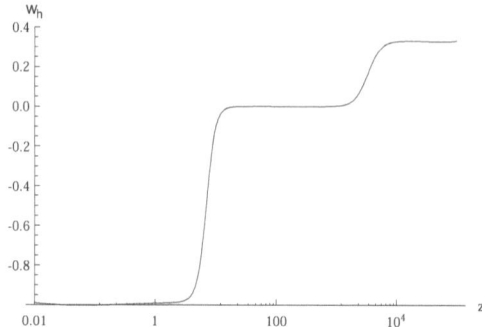

Figure 5.1: Equation of state of quintessence in the crossover quintessence model ($w_{h0} = -0.99, z_+ = 7$)

not building and solving models of this type in detail, but rather estimating general properties of the scalar evolution. Hence we will not start with appropriate potentials for crossover quintessence and evolve the cosmon over time, but rather simulate the behavior of the field by defining the quintessence equation of state by hand. We set the dark energy equation of state to constant at late times with value w_{h0}. Above some given redshift z_+ the equation of state crosses over to the scaling condition $w_h = 0$ in the matter dominated era; then for $z > z_{eq}$, before matter-radiation equality, we again have scaling through $w_h = 1/3$, where z_{eq} can be obtained via

$$z_{eq} = \frac{\Omega_M}{\Omega_\gamma} - 1 \,. \tag{5.18}$$

Then the general relation ($a = (1+z)^{-1}$)

$$\frac{d \ln \rho}{d \ln a} = -3(1 + w(a)) \tag{5.19}$$

may be used to find the matter, radiation and dark energy densities over cosmological time. Combining Eq. (5.15) with Eq. (5.11), we can estimate the scalar kinetic energy via

$$\dot{\varphi}^2/2 = \rho_h(1 + w_h)/2 \tag{5.20}$$

and thus integrate $d\varphi/da = \dot{\varphi}/aH$ from the present back to any previous redshift. The initial conditions are set by specifying the present densities of matter, radiation and dark energy and the model parameters w_{h0} and z_+. For illustration, we set $w_{h0} = -0.99$ and $z_+ = 7$. The resulting equation of state is displayed in Fig. 5.1, the corresponding evolution of energy components in Fig. 5.2 and the dimensionless cosmon field $\varphi/\bar{M}_P$ in Fig. 5.3. As can be seen from Fig. 5.3, in this type of models the scalar field has a monotonic evolution. Assuming a constant coupling δ to the fundamental varying parameter, usually α_X, the variation is given by

$$\Delta \ln \alpha_X(z) = \delta(\varphi(z) - \varphi(0)) \,. \tag{5.21}$$

Hence this ansatz implies a monotonic evolution of variations.

5.3. CROSSOVER QUINTESSENCE MODELS

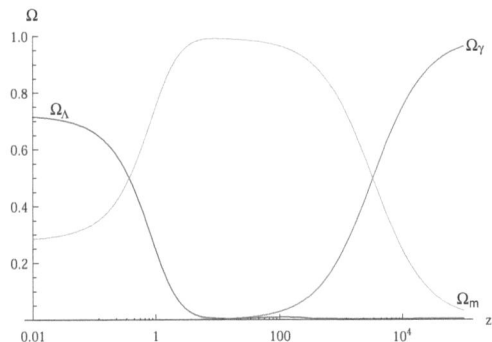

Figure 5.2: Energy components of our Universe in the crossover quintessence model $(w_{h0} = -0.99, z_+ = 7)$

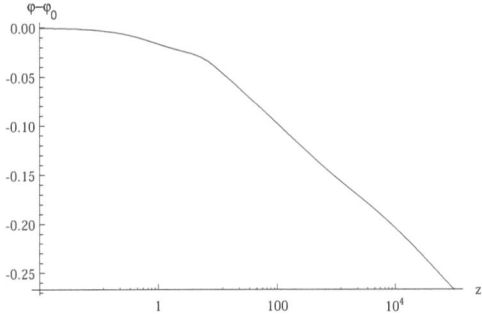

Figure 5.3: Dimensionless cosmon field φ in the crossover quintessence model $(w_{h0} = -0.99, z_+ = 7)$

5.4 Growing neutrino mass models

Growing neutrino models [Amendola08, Wetterich08] explain the value of today's dark energy density by the "principle of cosmological selection". The present fraction of dark energy, Ω_h^0, is set by a dynamical mechanism. The essential ingredient of this class of models is a neutrino mass that depends on the cosmon field φ and grows in the course of the cosmological evolution. As soon as the neutrinos become non-relativistic, their coupling to the cosmon triggers an effective stop (or substantial slowing) of the evolution of the cosmon. Before this event, the quintessence field follows the tracking behavior described in the preceding section. In the models which we will study, the cosmon is assumed to have the potential from Eq. (5.16),

$$V(\varphi) = \bar{M}_P^4 e^{-\alpha \frac{\varphi}{M_P}} .$$

The present dark energy density, ρ_{h0}, can be expressed in terms of the average present neutrino mass, $m_\nu(t_0)$, and a dimensionless parameter ζ of order unity [Amendola08],

$$(\rho_{h0})^{1/4} = 1.07 \left(\frac{\zeta m_\nu(t_0)}{eV} \right)^{1/4} 10^{-3} eV . \quad (5.22)$$

We follow again our simple proportionality assumption, namely that the cosmon coupling to a typical fundamental parameter is given by Eq. (5.21),

$$\Delta \ln \alpha_X(z) = \delta(\varphi(z) - \varphi(0)) ,$$

with a proportional variation for other couplings according to the unification scenario that we will study. This is the only contribution to the variation of the unified coupling α_X and M_{GUT}/M_P. However, a new ingredient is an additional variation of the Higgs v.e.v. $\langle \phi \rangle$ with respect to M_{GUT}, which only becomes relevant at late time [Wetterich08]. It is due to the effect of a changing weak triplet operator on the v.e.v. of the Higgs doublet. If the dominant contribution to the neutrino mass arises from the "cascade mechanism" (or "induced triplet mechanism") via the expectation value of this triplet, this changing triplet value is directly related to the growing neutrino mass [Wetterich08]. To understand this mechanism, we start with the most general mass matrix for the light neutrinos,

$$m_\nu = M_D M_R^{-1} M_D^T + M_L . \quad (5.23)$$

The first term is responsible for the seesaw mechanism [Minkowski77] with the mass matrix for heavy "right handed" neutrinos M_R and a Dirac mass term M_D. The second term accounts for the "induced triplet mechanism" [Magg80]

$$M_L \propto \frac{\langle \phi \rangle^2}{M_t^2} , \quad (5.24)$$

where a heavy $SU(2)_L$-triplet field t with mass M_t enters the equation (see [Wetterich08] for details). It is assumed that the mass of the triplet depends on the cosmon field, $M_t = M_t(\varphi)$.

The φ-dependence of the Higgs v.e.v. $\langle \phi \rangle$ is introduced by assuming a general effective potential $U(\varphi, \phi, t)$. Solving the field equations for the Higgs doublet field ϕ and the triplet field t, $\partial U/\partial \phi = 0$, $\partial U/\partial t = 0$, the cosmon potential is then obtained as

$$V(\varphi) = U(\varphi, \phi(\varphi), t(\varphi)) . \quad (5.25)$$

5.4. GROWING NEUTRINO MASS MODELS

In [Wetterich08] the simple potential

$$U = U_0(\varphi) + \frac{\lambda}{2}(\phi^2 - \phi_0^2)^2 + \frac{1}{2}M_t^2(\varphi)t^2 - \gamma\phi^2 t \tag{5.26}$$

is assumed, with γ and λ some coupling parameters. Solving the field equations for the Higgs doublet, it is found [Wetterich08]

$$\frac{\langle\phi\rangle^2}{M_{GUT}^2}(\varphi) = \frac{\langle\phi\rangle_0^2}{M_{GUT}^2}\left(1 - \frac{\gamma^2}{\lambda M_t^2(\varphi)}\right)^{-1}, \tag{5.27}$$

where $\langle\phi\rangle_0^2$ has to be chosen such that the measured Higgs v.e.v. is obtained today.

In the following we consider two models, with slightly different functional dependence of the Higgs v.e.v. and neutrino mass on the scalar field.

5.4.1 Stopping growing neutrino model

In the first model, the cosmon asymptotically approaches a constant value ("stopping growing neutrino model") [Wetterich08] and the neutrino mass is given by

$$m_\nu(\varphi) = \bar{m}_\nu\left\{1 - \exp\left[-\frac{\epsilon}{\bar{M}_P}(\varphi - \varphi_t)\right]\right\}^{-1}. \tag{5.28}$$

With a triplet mass dependence

$$M_t^2(\varphi) = \bar{M}_t^2\left[1 - \exp\left(-\frac{\epsilon}{\bar{M}_P}(\varphi - \varphi_t)\right)\right], \tag{5.29}$$

the additional Higgs variation is given according to

$$\frac{\langle\phi\rangle}{M_{GUT}}(z) = \bar{H}\left(1 - R(z)\right)^{-0.5}, \tag{5.30}$$

where

$$R(z) = \frac{R_0}{1 - \exp(-\frac{\epsilon}{\bar{M}_P}(\varphi(z) - \varphi_t))}. \tag{5.31}$$

Here, $\varphi_t \approx 27.6$ is the asymptotic value (choosing the parameter $\alpha = 10$ in the exponential potential [Wetterich08]). For illustration we take the set of parameters given in [Wetterich08], $\epsilon = -0.05$, and $\bar{H}$ is set by demanding the Higgs v.e.v. being consistent with measurements today, $\langle\phi\rangle(z=0) = 175\,\text{GeV}$. We set $R_0 = 10^{-7}$, however in general we only require $R(z=0) \ll 1$. The resulting variations are shown in Fig. 5.4 and Fig. 5.5.

The stopping growing neutrino model has an oscillation in $\langle\phi\rangle$ that grows both in frequency and amplitude at late times as φ approaches its asymptotic value. Such oscillations must not be too strong as measurements between $z = 2$ and today would measure a high rate of change. The oscillation may be made arbitrarily small by choosing small R_0. However, the linear variation (5.21) is independent of R_0.

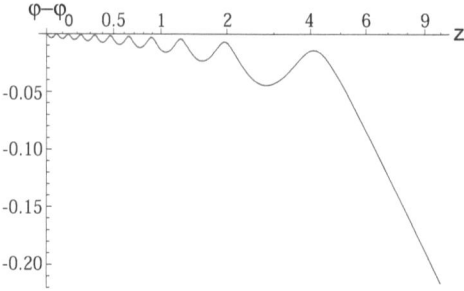

Figure 5.4: Evolution of the dimensionless cosmon field in the stopping growing neutrino model of [Wetterich08].

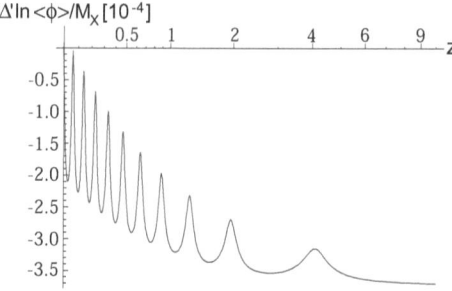

Figure 5.5: Additional variation of the Higgs v.e.v. according to Eq. (5.30) in the stopping growing neutrino model of [Wetterich08].

5.4. GROWING NEUTRINO MASS MODELS

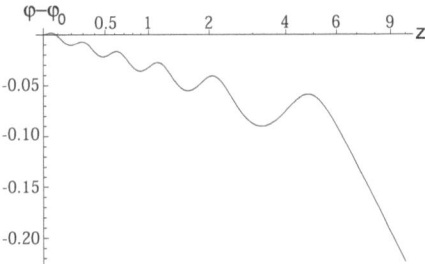

Figure 5.6: Evolution of the dimensionless cosmon field in the scaling growing neutrino model of [Amendola08].

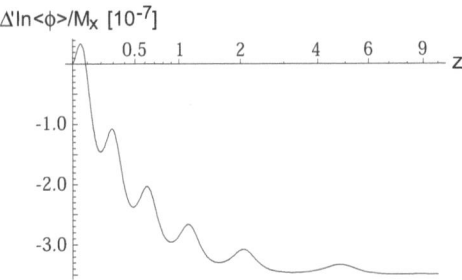

Figure 5.7: Additional variation of the Higgs v.e.v. according to Eq. (5.30) in the scaling growing neutrino model of [Amendola08].

5.4.2 Scaling growing neutrino model

The second growing neutrino model [Amendola08] does not lead to an asymptotically constant φ. Now the coupling of the neutrino to the cosmon φ is given by a constant β, according to

$$m_\nu = \tilde{m}_\nu e^{-\beta\varphi}. \tag{5.32}$$

This "scaling growing neutrino model" leads in the future to a scaling solution with a constant ratio between the neutrino and cosmon contributions to the energy density.

With the choice of parameters $\beta = -52$, $\alpha = 10$ and $m_{\nu,0} = 2.3\,\text{eV}$ [Amendola08], and given the triplet mechanism of [Wetterich08], the Higgs v.e.v. varies as Eq. (5.30), where now R is given by

$$R(z) = R_0 e^{-\beta\varphi(z)}. \tag{5.33}$$

Here the Higgs oscillations remain comparatively small in amplitude, while the absolute value of $\langle \phi \rangle$ grows overall with time: see Figs. 5.6 and 5.7 with the parameter choice $R_0 = 10^{-6}$. Compared to the "stopping growing neutrino model", this model has milder oscillations at late time.

5.5 A short note on string theory

String theory is a candidate for a "theory of everything". The underlying concept is that fundamental particles are one dimensional objects (strings) which live in a 10 dimensional spacetime. As the world as we know it has only four spacetime dimensions, the remaining 6 dimensions are usually compactified in a way that only 4 "large" spacetime dimensions remain. During compactification, several scalar fields appear which can in principle have all sorts of couplings to the Standard Model fields. A further scalar field, the dilaton, is present in string theory from the very beginning. Its v.e.v. sets the string theory coupling constant.

With a multitude of possible fields and couplings, one should in principle be able to model any quintessence scenario within the framework of string theory. However, there are strong arguments against it, in particular based on the small mass and potential of the quintessence field (see *e.g.* [Banks01]). A deeper study of variations implied by string theory or possible tests of string theory is out of scope of this thesis, but in principle our methods will also apply to any string theory induced variations of constants, as long as they can be described with effective field theories. Then our methods also allow to constrain the allowed regions in the landscape of string theory.

Part II

Big Bang Nucleosynthesis

Chapter 6

Big Bang Nucleosynthesis

6.1 Why BBN?

As we have mentioned in Sec. 2.3, changes in fundamental constants are most likely to appear over large time scales and / or different environments. Big Bang Nucleosynthesis is the earliest process in the history of the Universe which can be both reasonably described with standard physics and astrophysically probed. Also, the Universe was much denser at this time[1], even though locally the environment one finds for instance in heavy stars, supernovae or black holes is much more extreme than at BBN. Hence, it is very reasonable to carefully study the influence of varying parameters on the process of primordial element production.

6.2 How will we study BBN?

BBN is a complex process involving a lot of nuclear reactions and particles. There are two main approaches to a theoretical prediction of primordial element abundances. The first one is purely analytical and only gives rough estimates, whereas the second one numerically simulates the whole process of BBN and gives high-precision abundance predictions.

For our purpose, analytical estimates will turn out to have insufficient accuracy, hence we will utilize a numerical procedure to obtain our findings on the influence of varying constants on BBN. Numerical codes for the simulation of primordial nucleosynthesis go back to the late 60s and 70s [Peebles66, Wagoner69, Wagoner72]. Our BBN code is based on the Kawano 1992 code [Kawano88, Kawano92], with updated nuclear reaction cross sections as given in [NACRE99] and [NETGEN]. We have improved the code in terms of numerical accuracy in order to be able to derive high precision parameter dependences (see Sec. 6.7.1 for details). In the following sections we explain the process of BBN and the underlying physics in greater detail. As has been explained in Sec. 2.5, we will work in a system of units where Λ_{QCD} is kept constant.

---

[1]The energy density at BBN was roughly the same as the the density of normal water.

6.3. THE PROCESS OF BBN

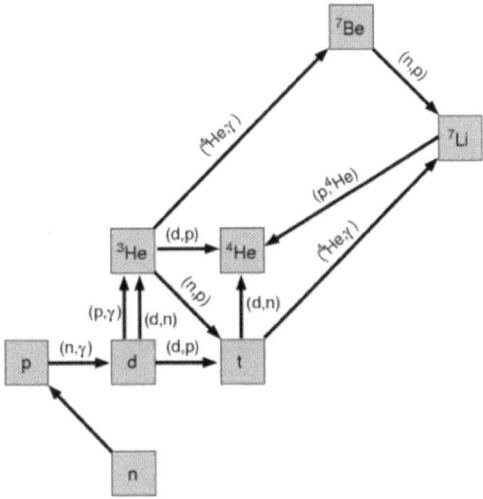

Figure 6.1: Network of main reactions responsible for primordial nucleosynthesis. From [Bartelmann06].

6.3 The process of BBN

In the standard BBN process (SBBN), the element synthesis does not depend on any pre-BBN phase. At a temperature of, for instance, $T = 10^{11}$K, baryogenesis and quark condensation have ended and the Universe only contains protons and neutrons in the baryonic sector. These are held in equilibrium via the weak reactions

$$p + e^- \longleftrightarrow n + \nu_e$$
$$n + e^+ \longleftrightarrow p + \bar{\nu}_e. \quad (6.1)$$

As the Universe expands and cools down, these reactions freeze out at $T \approx 800$ keV, and the neutrons decay freely. The next step in nucleosynthesis is the fusion of deuterium, which happens when reaction

$$n + p \longleftrightarrow D + \gamma$$

drops out of equilibrium. Due to the large photodissociation cross section, any produced deuterium is immediately destroyed by photons from the background radiation until their temperature has dropped considerably below the binding energy of deuterium, $T \approx B_D = 2.2$ MeV. In fact, since photons are much more abundant than baryons, the high-energy photons in the Maxwell tail keep the reaction in equilibrium until the temperature reaches $T \approx 80$ keV. Once deuterium fusion sets in, elements up to mass number A=7 are synthesized via a network of 11 main nuclear reactions displayed in Fig. 6.1. As the Universe continues expanding and cooling down, the temperature and density will fall below the level required for the nuclear reactions at some point. This happens a few minutes after the Big Bang, when the reaction rates

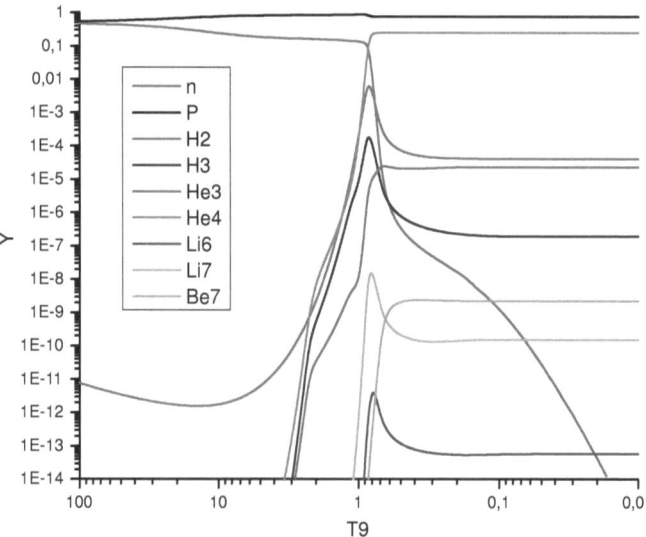

Figure 6.2: Element abundances Y as a function of the decreasing temperature of the universe $T_9 = T \times 10^{-9} \text{K}^{-1}$

become slower than the expansion rate of the Universe,

$$n\langle \sigma v \rangle < H ,\qquad(6.2)$$

with n the particle number density and $\langle \sigma v \rangle$ the Maxwell-Boltzmann averaged cross section (see Sec. 6.4.3). At the end of the BBN process, the element composition of the Universe is roughly 75% hydrogen, 25% helium (numbers w.r.t. mass) and small amounts of deuterium and ^{7}Li. Two elements produced during BBN, ^{7}Be and tritium, decay via a slow β-decay to ^{7}Li and ^{3}He respectively after BBN has ended. In fact, the primordial ^{7}Li abundance primarily derives from ^{7}Be produced during BBN.

The evolution of element abundances during the process of BBN is shown in Fig. 6.2, where we define the mass abundance Y_i for a nucleus i as

$$Y_i := A_i \frac{n_i}{n_B} \qquad(6.3)$$

with n_B the baryon number density, n_i the number density of nucleus i and A_i its mass number.

Beyond $A = 8$ elements

During primordial nucleosynthesis, essentially no elements are produced beyond the $A = 7$ mass limit. Heavier elements, including carbon and oxygen which are ubiquitous on earth, have been produced later in stars and were thrown back into space during supernova explosions. The process leading to those elements proceeds over the

6.4. THE PHYSICS OF BBN

unstable nucleus ^{8}Be via the "triple-alpha process"

$$^4\text{He} + {}^4\text{He} \rightarrow {}^8\text{Be} + \gamma$$
$$^8\text{Be} + {}^4\text{He} \rightarrow {}^{12}\text{C} + \gamma \,.$$

Due to the low density and low temperature at BBN (compared to stars), the probability that a ^{8}Be nucleus meets a ^{4}He nucleus during its lifetime of 67×10^{-18}s is extremely low. Hence, the absence of stable $A = 8$ nuclei prohibits any element fusion beyond ^{7}Li during BBN.

6.4 The physics of BBN

In this section we will describe the underlying processes of BBN in greater detail and introduce the relevant equations.

6.4.1 Cosmological background equations

BBN is described in the framework of the FRW metric. The basic equations describing the expansion of the Universe are the two Friedmann equations (3.4) and (3.5). The process of BBN happened at the a redshift of $z \sim 10^{10}$, a time when the Universe was radiation dominated (see Fig. 3.1), *i.e.* the expansion rate of the Universe is completely controlled by relativistic particles. The temperature T of the Universe is defined by the photon temperature T_γ, and the Stefan-Boltzmann law can be used to express energy and number density of the photons in terms of T_γ,

$$\rho_\gamma = \frac{\pi}{15} \frac{k_B^4}{(c\hbar)^3} T_\gamma^4 \tag{6.4}$$

$$n_\gamma = \frac{2\zeta(3)}{\pi^2} \left(\frac{k_B T_\gamma}{\hbar c}\right)^3 \,. \tag{6.5}$$

BBN is taking place in the temperature regime 3×10^9K $<$ T $< 0.1 \times 10^9$K. Combining Eq. (6.4) with the Friedmann equation (3.4) yields a relation between temperature and time after the Big Bang,

$$t = \left(\frac{16\pi^3 G_N}{5\hbar^3 c^5}\right)^{-1/2} (k_B T)^{-2} \,. \tag{6.6}$$

This relation can be applied in the early Universe until the number of photons in the Universe is increased by $e^+ e^-$ annihilation.

Before BBN, neutrinos, photons and electrons were in equilibrium, for instance via

$$\nu + \bar{\nu} \longleftrightarrow e^+ + e^- \longleftrightarrow 2\gamma \,. \tag{6.7}$$

This reaction freezes out at $T \approx 1.5$ MeV [Fornengo97], but as also the neutrinos are ultra relativistic, both T_γ and T_ν depend on the scale factor[2]

$$T \sim \frac{1}{a} \tag{6.8}$$

---
[2]Equation (6.8) also gives a relation between temperature and redshift, $T \sim (1+z)$. Note however that this relation does not hold during reheating.

and consequently should evolve equally over time. Due to e^+e^- annihilation, the photon temperature increases whilst the neutrino temperature strictly follows the $T \sim \frac{1}{a}$ law. With T_γ and T_ν evolving differently, the two temperatures have to be tracked separately during BBN. For the neutrino temperature we get an expression equivalent to the Stefan-Boltzmann law,

$$\rho_\nu = N_\nu \frac{7}{8} \frac{\pi}{15} \frac{k_B^4}{(c\hbar)^3} T_\nu^4 , \qquad (6.9)$$

where N_ν is the number of neutrino generations (in SBBN $N_\nu = 3$).

Note that temperatures can be given both in units of Kelvin (K) and energy (MeV) via $E = k_B T$, with the conversion factor

$$1 \text{ MeV} \simeq 11.6 \cdot 10^9 \text{K} \qquad (6.10)$$

e^+e^- annihilation

At $T \approx 10^{10}$ K ($E \approx 1$ MeV) the reaction $e^+ + e^- \leftrightarrow 2\gamma$ drops out of thermodynamic equilibrium. Electrons and positrons annihilate to photons, the number of photons and hence also the temperature of the photon gas rises (this is called "reheating"). Entropy conservation yields that the number of photons, and hence the photon energy- and number density, rise by a factor $11/4$. This is also the reason for today's difference in photon and neutrino temperature, $T_\gamma^0 = 2.73$ K, $T_\nu^0 = 1.95$ K ($T_\gamma^0/T_\nu^0 = (11/4)^{1/3}$).

6.4.2 Initial conditions

We start our simulation of the BBN process at a temperature of $T = 10^{11}$ K. As has been explained in Sec. 6.3, protons and neutrons are in thermodynamic equilibrium via the reaction (6.1), hence the neutron-to-proton ratio is given by the thermodynamic relation

$$\frac{n}{p} = e^{-(m_n - m_p)c^2/k_B T} . \qquad (6.11)$$

Knowing the starting temperature (which has to be well above the freezeout temperature of the reaction (6.1)), the initial ratio of neutrons to protons is well defined. Via Eq. (6.6) the initial temperature can be translated into the initial time t_{init}.

The only additional cosmological parameter necessary for BBN is the initial baryon number density[3],

$$n_B^{initial} = n_n^{initial} + n_p^{initial} . \qquad (6.12)$$

In principle, this parameter should drop out of a theory of baryogenesis. However, lacking this theory, we have to take n_B as a cosmological parameter which has to be plugged into the BBN simulation by hand. Since n_B scales like the photon number density n_γ during cosmological expansion,

$$n_B \propto n_\gamma \propto a^{-3} , \qquad (6.13)$$

the parameter

$$\eta := n_B/n_\gamma = 2.74 \cdot 10^{-8} \Omega_b h^2 \qquad (6.14)$$

---

[3] A further cosmological parameter often considered in BBN studies is the 'number of neutrino species' at BBN, *i.e.* the number of relativistic particles present at BBN. Since the presence of those additional particles is equivalent to a change in expansion rate, we treat this parameter effectively as a change in the gravitational constant G_N. In SBBN, η is the only parameter which enters, and N_ν is set to 3.

6.4. THE PHYSICS OF BBN

stays constant throughout the later evolution of the Universe, unless the number density of the photons is not significantly increased by later events, e.g. by the decay of particles. This is exactly what happens during e^+e^- annhililation as described in section 6.4.1. The effect on the photon temperature could be computed, giving that η at $T > 10^{10}$ K was higher than the current η_0 by a factor $\frac{11}{4}$, $\eta_{T>10^{10}K} = \frac{11}{4}\eta_0$. Including this correction factor, we can directly deduce the baryon density at BBN (for a given temperature $T \approx 10^9$ K) from today's value of η by applying Eq. (6.5),

$$n_B^{BBN} = \frac{11}{4}\eta_0 \cdot \frac{2\zeta(3)}{\pi^2}\left(\frac{k_B T}{\hbar c}\right)^3. \tag{6.15}$$

As this quantity determines the frequency of nuclear collisions, it is clear that a modified n_B will change the "speed" of the BBN process.

6.4.3 The element synthesis process

During BBN, element production is dominantly driven by three types of processes[4],

- $2 \to 1$ fusion processes $A + B \to C$,
- $2 \to 2$ scatterings $A + B \to C + D$, and
- particle decays, $A \to B + C$.

The parameters describing these processes are the reaction cross sections $\sigma_{A+B \to C}$, $\sigma_{A+B \to C+D}$ and the decay width $\lambda_{A \to B+C} \equiv \tau^{-1}$ (inverse mean lifetime). Given these quantities, one derives simple differential equations that describe the time evolution of the number densities of the nuclei A, B, C, D, n_A, n_B, n_C, n_D.

- For a decay $A \to B + C$,

$$\frac{dn_A}{dt} = -\lambda n_A \tag{6.16}$$

$$\frac{dn_B}{dt} = +\lambda n_A \tag{6.17}$$

$$\frac{dn_C}{dt} = +\lambda n_A \tag{6.18}$$

- For a $2 \to 1$ process $A + B \to C$

$$\frac{dn_A}{dt} = -n_A n_B \langle \sigma v \rangle \tag{6.19}$$

$$\frac{dn_B}{dt} = -n_A n_B \langle \sigma v \rangle \tag{6.20}$$

$$\frac{dn_C}{dt} = +n_A n_B \langle \sigma v \rangle \tag{6.21}$$

---

[4]Due to the comparably low temperature and density at BBN, three-particle interactions which are important in the stellar nucleosynthesis process do not play a role at BBN [Aprahamian05] and are hence not considered in BBN simulations.

- For a $2 \to 2$ process $A + B \to C + D$

$$\frac{dn_A}{dt} = -n_A n_B \langle \sigma v \rangle \qquad (6.22)$$

$$\frac{dn_B}{dt} = -n_A n_B \langle \sigma v \rangle \qquad (6.23)$$

$$\frac{dn_C}{dt} = +n_A n_B \langle \sigma v \rangle \qquad (6.24)$$

$$\frac{dn_D}{dt} = +n_A n_B \langle \sigma v \rangle \,. \qquad (6.25)$$

Here $\langle \sigma v \rangle$ is the Maxwell-Boltzmann averaged cross section,

$$\langle \sigma v \rangle = \frac{(8/\pi)^{1/2}}{\mu^{1/2}(k_B T)^{3/2}} \int_0^\infty \sigma(E) E \exp(-E/k_B T) dE \,, \qquad (6.26)$$

where μ is the reduced mass of the reactants, and E is the reaction energy in the center of mass system.

6.5 Nuclear reaction rates and the Q value

For nuclear reactions, one of the most important quantities is the reaction Q value, which gives the amount of energy released (or absorbed) in a reaction. Working with positive binding energies[5] B_i, the Q value for a reaction is defined as

$$Q := \sum_{outgoing} B_{out} - \sum_{incoming} B_{in} \,. \qquad (6.27)$$

Hence each reaction Q-value is determined by the masses of reactants and products. The Q-value of each reaction mainly affects the abundances via the reverse thermal reaction rate relative to the forward rate, and via phase space and radiative emission factors in the reaction cross-sections.

The reverse reaction rate is simply related to the forward rate via statistical factors, due to time reversal invariance (see for example [NACRE99]): the relevant dependence for a $12 \to 34$ reaction is

$$\frac{\langle \sigma v \rangle_{34 \to 12}}{\langle \sigma v \rangle_{12 \to 34}} \propto e^{-Q/T} \,. \qquad (6.28)$$

The Q-dependence of radiative capture reactions $A + B \to C + \gamma$, assuming a dominant electric dipole, is

$$\sigma(E) \propto E_\gamma^3 \sim (Q + E)^3 \,, \qquad (6.29)$$

whereas for $2 \to 2$ inelastic scattering or transfer reactions the dependence is

$$\sigma(E) \propto \beta \sim (Q + E)^{1/2} \,, \qquad (6.30)$$

where β is the outgoing channel velocity. In the current treatment we assume $E \ll Q$ at relevant temperatures, and simply scale rates by the appropriate power of Q.[6]

---

[5] The positive binding energy is defined via $B(A, Z) := Z m_p + (A - Z) m_n - m(A, Z)$.

[6] Clearly this breaks down when Q approaches zero, and we have not considered varying any binding energy to the point where this happens. A more accurate treatment would involve applying the phase space dependences directly to the cross-sections, for example in the S-factor description (see Sec. 6.6.3) of charged particle reactions, which involves an expansion in E; the dependence on $(Q + E)$ can then be applied order by order.

6.6. NUCLEAR REACTIONS IMPORTANT FOR BBN

Reaction	Q value [MeV]	D	^{3}He	^{4}He	^{6}Li	^{7}Li
$n \leftrightarrow p$	1.29	-0.80	-0.30	-1.48	-2.76	-0.92
$p(n,\gamma)d$	2.22	-0.2	0.1	0	-0.2	1.3
$d(p,\gamma)^3$He	5.49	-0.3	0.4	0	-0.3	0.6
$d(d,n)^3$He	3.27	-0.5	0.2	0	-0.5	0.7
$d(d,p)t$	4.03	-0.5	-0.3	0	-0.5	0.1
$d(\alpha,\gamma)^6$Li	1.47	0	0	0	1.0	0
^{3}He$(n,p)t$	0.76	0	-0.2	0	0	-0.3
^{3}He$(d,p)^4$He	18.35	0	-0.8	0	0	-0.7
^{3}He$(\alpha,\gamma)^7$Be	1.59	0	0	0	0	1.0
^{6}Li$(p,\alpha)^3$He	4.02	0	0	0	-1.0	0
^{7}Be$(n,p)^7$Li	1.64	0	0	0	0	-0.7
^{7}Be(n,a)^{4}He	18.99	0	0	0	0	-0.01
T(p,g)^{4}He	19.81	0	0	0	0	0.02
^{7}Li(p,a)^{4}He	17.35	0	0	0	0	-0.06
T(a,g)^{7}Li	2.47	0	0	0	0	0.03
T(d,n)^{4}He	17.59	0	-0.01	0	0	-0.02
^{7}Be(d,pa)^{4}He	16.77	0	0	0	0	-0.01

Table 6.1: Leading dependence of abundances on thermal averaged cross-sections $\partial \ln Y_a / \partial \ln \langle \sigma v \rangle_i$ for important reactions (1st part) and less important reactions (2nd part)

A variation in binding energies can have two kinds of effect. The Q value can change the time when a reaction drops out of equilibrium, for instance at the $n+p \to d+\gamma$ reaction. Or it can change the absolute rate of a reaction, and thus the production rate of a given species, for example at the ^{7}Be-producing reaction whose cross-section varies with Q^3.

Whether the reaction matrix elements have a dependence on the binding energies and on Q is in general not clear because there is no systematic effective theory for multi-nucleon reactions. The exception is the $np \to d\gamma$ reaction, for which we have implemented a nuclear effective theory result (see Sec. 6.6.2).

6.6 Nuclear reactions important for BBN

Simulations of the BBN process usually track a large amount of elements and reactions. However, only a few of them turn out to be important for BBN. In the next section, we will be interested in the behavior of BBN under variations of parameters, hence we will be concentrating on those reactions where a variation of parameters will result in a variation in final abundances. In order to estimate which reactions are more or less important for the variation of the final abundances, we varied each thermal averaged cross-section $\langle \sigma v \rangle$ by a temperature-independent factor, preserving the relation between forward and reverse rates. The resulting values for $\partial \ln Y_a / \partial \ln \langle \sigma v \rangle_i$ are given in Tab. 6.1. As usual in BBN simulations, the slow β-decays of tritium and ^{7}Be are accounted for by adding on the T and ^{7}Be abundances to ^{3}He and ^{7}Li respectively at the end of the run, when other nuclear reactions have frozen out.

For all reactions in Tab. 6.1, the comments on Q-value (and hence binding energy) dependence of the preceding sections do apply. In the following sections we will give details on those reactions for which a more detailed dependence on fundamental parameters is known.

6.6.1 The $n \leftrightarrow p$ reaction rate

The $n \leftrightarrow p$ weak interactions influence every abundance nontrivially. Opposed to most of the reaction rates which can only be determined experimentally, a closed formula is known for the $n \leftrightarrow p$ weak reaction [Scherrer83, Lopez97, WeinbergGRT].

$$\lambda_{n \to p} = \lambda_{en \to \nu p} + \lambda_{\nu n \to ep} + \lambda_{n \to pe\nu} \tag{6.31}$$

$$\lambda_{p \to n} = \lambda_{ep \to \nu n} + \lambda_{\nu p \to en} + \lambda_{pe\nu \to n} \tag{6.32}$$

$$\lambda_{n \to pe\nu} = K \int_1^q d\epsilon \frac{(\epsilon - q)^2 (\epsilon^2 - 1)^{1/2} \epsilon}{[1 + \exp(-\epsilon z)][1 + \exp((\epsilon - q)z_\nu - \xi_e])} \tag{6.33}$$

$$\lambda_{n\nu \to pe} = K \int_q^\infty d\epsilon \frac{(\epsilon - q)^2 (\epsilon^2 - 1)^{1/2} \epsilon}{[1 + \exp(-\epsilon z)][1 + \exp((\epsilon - q)z_\nu - \xi_e])} \tag{6.34}$$

$$\lambda_{ne \to p\nu} = K \int_1^\infty d\epsilon \frac{(\epsilon + q)^2 (\epsilon^2 - 1)^{1/2} \epsilon}{[1 + \exp(\epsilon z)][1 + \exp(-(\epsilon + q)z_\nu - \xi_e])} . \tag{6.35}$$

Here $q := (m_n - m_p)/m_e$, $z = m_e/T_\gamma$, $z_\nu = m_e/T_\nu$ and ξ_e is the electron neutrino degeneracy parameter which we always set to zero. The constant K is obtained by demanding $\lambda_{n \to p}(T \to 0) \equiv \tau_n^{-1}$. The corresponding $p \to n$ rates are obtained via $\lambda_{p \to n} = \lambda_{n \to p}(-q, -\xi_e)$.

The equations given above derive from first-order electroweak theory and contain some approximations to allow an easy numerical evaluation. However, higher precision of measurements demands a higher accuracy for theoretical simulations. Hence, modern simulations of BBN apply corrections to the equations given above. Those are radiative corrections in the form of zero-temperature and thermal Coulomb corrections [Lopez98], as well as finite nucleon mass corrections given in [Lopez97]. For example, thermal Coulomb corrections are applied by multiplying the integrand of the rates for $n \leftrightarrow pe\nu$ and $ep \leftrightarrow n\nu$ with the Fermi factor

$$\frac{2\pi\alpha/\beta}{1 - \exp(-2\pi\alpha/\beta)} , \tag{6.36}$$

where $\beta = \sqrt{1 - \epsilon^{-2}}$. Similar prescriptions exist for the other corrections [Lopez97, Lopez98].

6.6. NUCLEAR REACTIONS IMPORTANT FOR BBN

6.6.2 The $n + p \to D + \gamma$ reaction rate

Chen *et al.* derived an effective theory for the strong $n + p \to D + \gamma$ cross section which is now widely used [Chen99]. They use an effective theory which contains direct nucleon-nucleon interactions and photon exchange, but neglects pion exchange[7]. They estimate the error in the energy regime of BBN to be $\lesssim 4\%$. Using nucleon-nucleon phase-shift data and the cross section for cold neutron capture as input data, they obtain the folloing equations

$$\sigma_{np \to d\gamma} = \frac{4\pi\alpha(\gamma^2 + P^2)^3}{\gamma^3 m_N^4 P} \left(\tilde{X}_{M1}^2 + \tilde{X}_{E1}^2 \right), \tag{6.37}$$

where $P = \sqrt{m_N E}$, $\gamma = \sqrt{B_D m_N}$, B_D is the deuteron binding energy and E is the cross section energy in the center of mass system. $\tilde{X}_{M1}^2$ and $\tilde{X}_{E1}^2$ are given by

$$\tilde{X}_{E1}^2 = \frac{P^2 m_N^2 \gamma^4}{(\gamma^2 + P^2)^4} \left[1 + \gamma\rho_d + (\gamma\rho_d)^2 + (\gamma\rho_d)^3 + \frac{m_N \gamma}{6\pi} \left(\frac{\gamma^2}{3} + P^2 \right) C_1 \right], \tag{6.38}$$

$$\tilde{X}_{M1}^2 = \frac{\kappa_1^2 \gamma^4 \left(\frac{1}{a_1} - \gamma \right)^2}{\left(\frac{1}{a_1^2} + P^2 \right)(\gamma^2 + P^2)^2} \left[1 + \gamma\rho_d - r_0 \frac{\left(\frac{\gamma}{a_1} + P^2 \right) P^2}{\left(\frac{1}{a_1^2} + P^2 \right)\left(\frac{1}{a_1} - \gamma \right)} - \frac{L_{np} m_N}{2\pi \kappa_1} \frac{\gamma^2 + P^2}{\frac{1}{a_1} - \gamma} \right], \tag{6.39}$$

with the EFT fitting constants [Chen98, Chen99]

ρ_d	=	1.764 fm	(effective range parameter)
C_1	=	-1.49 fm^4	(P-wave interaction constant)
κ_1	=	2.352945	(isovector nucleon magnetic moment)
a_1	=	-23.714 fm	(scattering length in the $^1S_0$ channel)
r_0	=	2.73 fm	(effective range in the $^1S_0$ channel)
L_{np}	=	-4.513 fm^2	($M1$ capture constant) .

6.6.3 Charged particle reaction rates

Theoretical models describing the rough shape of charged particle reaction cross sections have long been available [Fowler67]. In absence of resonances, the cross sections are a product of the Gamow factor and an "S-factor",

$$\sigma(E) = S(E) \frac{e^{-2\pi \tilde{\eta}}}{E}, \tag{6.40}$$

where

$$\tilde{\eta} \equiv \alpha Z_1 Z_2 \sqrt{\mu/2E}, \tag{6.41}$$

$Z_{1,2}$ the atomic numbers in the initial state and μ is the reduced mass. The S-factor may be expanded in a Maclaurin series to quadratic order in energy, which is usually sufficient to account for any smoothly-varying dependence,

$$S(E) = S_0 + S_1 E + S_2 E^2. \tag{6.42}$$

---

[7] In EFT treatments of processes where the external momenta are much smaller than the pion mass (as is the case at BBN with nucleon energies $E_N \lesssim 1$ MeV, $m_\pi \approx 135$ MeV), a pion-less EFT turns out to describe the process sufficiently well [Gegelia98, ChenRupak99].

To obtain the Maxwell-Boltzmann averaged cross section (Eq. (6.26)) one has to perform an integration over energy, resulting in a sum of terms for $\langle \sigma v \rangle$ with specific dependence on α and the reduced mass μ. A generic cross section has the form [Bergstrom99][8]

$$\langle \sigma v \rangle = a_1 T_9^{-\frac{2}{3}} \left(\frac{\alpha}{\alpha_0}\right)^{\frac{1}{3}} \left(\frac{\mu}{\mu_0}\right)^{-\frac{1}{3}} e^{-3\kappa^{1/3}}$$
$$\times \left[1 + \frac{5}{36}\kappa^{-\frac{1}{3}} + a_2 T_9^{\frac{2}{3}} \left(\frac{\alpha}{\alpha_0}\right)^{\frac{2}{3}} \left(\frac{\mu}{\mu_0}\right)^{\frac{1}{3}} \cdot \left(1 + \frac{35}{36}\kappa^{-\frac{1}{3}}\right) \right.$$
$$\left. + a_3 T_9^{\frac{4}{3}} \left(\frac{\alpha}{\alpha_0}\right)^{\frac{4}{3}} \left(\frac{\mu}{\mu_0}\right)^{\frac{2}{3}} \cdot \left(1 + \frac{89}{36}\kappa^{-\frac{1}{3}}\right)\right] + [\text{resonance terms}], \quad (6.43)$$

where

$$\kappa := \pi^2 \alpha^2 Z_1^2 Z_2^2 \frac{\mu}{2k_B T} \quad (6.44)$$

and the a_i are fitting constants which are fit to the measured cross-sections. Some cross-sections are fit with an additional exponential term $\tilde{S}(0)e^{-\beta E}$ [Fowler75]. In addition, non-resonant terms may be multiplied by a cutoff factor $f_{\text{cut}} = e^{-(T/T_{\text{cut}})^2}$, where T_{cut} has been argued to be proportional to α^{-1} [Bergstrom99, Fowler75].

Resonances

Where the cross-section as a function of energy shows one or more resonances, they contribute to the thermal averaged rate as

$$\langle \sigma v \rangle_{\text{res}} = g(T) e^{-\bar{E}/T} , \quad (6.45)$$

where $g(T)$ and $\bar{E}$ are fitting parameters corresponding to the shape and position of the resonance. Usually a power-law is taken for $g(T)$, thus $g(T) = cT^p$. In principle one should consider the variation of the resonance parameters if this term is significant. But since the major contribution to the resonance energy $\bar{E}$ probably arises from Λ_{QCD} which we take as our (non-varying) unit, it seems a reasonable first guess to keep the resonance parameters fixed.

For the code, the NACRE formulae [NACRE99] fitted at the level of the thermal averaged cross-sections $\langle \sigma v \rangle$ are not suitable as they do not allow to incorporate the dependence on physical parameters. We use the functional forms of rates from [Smith92, Bergstrom99] which have been described above in Eq. (6.43) and fit the free parameters of these rates to reproduce the NETGEN rates [NETGEN] as closely as possible. We also checked that the resulting cross-sections are consistent with experimental data. In the case of $d(\alpha,\gamma)^6$Li we found a set of parameters which seems to fit the experimental cross-section at low energies [Kiener6Li] better than the NACRE fit. But note that this cross-section is not measured directly, rather it is derived from experimental data under various assumptions.

---

[8][Bergstrom99] only gives the dependence on α, we have recomputed the expansion including the dependence on the reduced mass μ.

Replacing the NETGEN rates in the code with our fitted rates, the obtained abundances change as follows:

- Y_D differs by -0.3%,
- Y_{3He} by $+0.9\%$,
- Y_{4He} by less than 0.1%, and
- Y_{7Li} by $+3\%$.

Hence we do not consider this refitting as significant, except in the case of the $d(\alpha,\gamma)^6Li$ reaction. Depending on whether this reaction is fit to NETGEN, or to the cross-section values of [Kiener6Li], we found a ^6Li abundance larger by a factor of 1.02, or 3.3, respectively. However, given the unclear observational status of ^6Li this discrepancy is not currently worth pursuing (see Sec. 6.8.5).

6.7 The simulation of the BBN process

The BBN process has been described and simulated more than 40 years ago [Wagoner66]. One starts with the initial conditions given in Sec. 6.4.2 and evolves the element synthesis processes for a set of nuclei as described by the differential equations in Sec. 6.4.3 in the background of an expanding universe. As we are working at a time when the Universe was radiation-dominated, the expansion of the Universe as given by the Friedmann equation (3.4) is dominated by the photon, neutrino and electron contributions[9]. Hence the energy density ρ entering in Eq. (3.4) is

$$\rho = \rho_\gamma + \rho_\nu + \rho_e \tag{6.46}$$

and the corresponding pressure

$$p = \frac{1}{3}\rho_\gamma c^2 + \frac{1}{3}\rho_\nu c^2 + p_e \,. \tag{6.47}$$

Whilst photons and neutrinos are ultra relativistic throughout BBN, the electrons move from the relativistic to the non relativistic domain in the course of the BBN simulation (10 MeV $\gtrsim T \gtrsim$ 0.001 MeV). Hence the electron density and pressure have to be described very carefully.

For the electron energy and pressure density one introduces the electron chemical potential

$$\phi_e \sim \frac{\pi^2 (\hbar c)^3 n_B Y_p}{2(k_B T z)^3} \left[\frac{1}{\sum_n (-)^{n+1} n L(nz)} \right] , \tag{6.48}$$

where $z = \frac{m_e c^2}{k_B T}$. The electron and positron energy and pressure densities are then [FowlerHoyle64]

$$\rho_e \equiv \rho_{e^-} + \rho_{e^+} = \frac{2(m_e c^2)^4}{\pi^2 (\hbar c)^3} \sum_{n=1}^{\infty} (-)^{n+1} \cosh(n\phi_e) M(nz) \tag{6.49}$$

$$\frac{p_{e^-} + p_{e^+}}{c^2} = \frac{2(m_e c^2)^4}{\pi^2 (\hbar c)^3} \sum_{n=1}^{\infty} \frac{(-)^{n+1}}{nz} \cosh(n\phi_e) L(nz) \,. \tag{6.50}$$

---

[9] The computer code also includes the evolution of baryonic and dark matter. However, this does not make any impact on the obtained abundances.

The functions L and M are related to the modified Bessel functions K_i via

$$L(z) := K_2(z)/z \tag{6.51}$$

$$M(z) := \frac{1}{z}\left[\frac{3}{4}K_3(z) + \frac{1}{4}K_1(z)\right]. \tag{6.52}$$

In our BBN simulation, we evolve the following set of parameters with time

- temperatures T_γ, T_ν
- electron chemical potential ϕ_e
- baryon density ρ_B (or equivalently the scale factor a)
- the abundances Y_i

according to the differential equations given in this chapter, and the Friedmann equation (3.4) is used to evolve the baryon density and temperatures according to the expansion of the Universe.

6.7.1 Numerical aspects of the BBN simulation

The main task of a BBN code is to numerically integrate the set of coupled differential equations which have been given in the preceding sections. In our code, the differential equations are solved using an adaptive second-order Runge-Kutta method, which is important to correctly account for the nuclear reactions[10]. Given a differential equation

$$y'(t) = f(t, y) \tag{6.53}$$

and a starting value at time t_0, $y(t_0) = y_0$, the task is to derive the value of y at time $t_0 + \Delta t$. The second order Runge-Kutta method does this in a two-step approach. First one derives the value in linear approximation, *i.e.*

$$\tilde{y}(t_o + \Delta t) = y(t_0) + \Delta t \cdot y'(t_0). \tag{6.54}$$

The "average" derivative between t_0 and $t_0 + \Delta t$ is then defined as

$$\tilde{y}' := \frac{f(t_0, y_0) + f(t_0 + \Delta t, \tilde{y}(t_0 + \Delta t))}{2} \tag{6.55}$$

and the final value

$$y(t_0 + \Delta t) = y_0 + \tilde{y}' \cdot \Delta t. \tag{6.56}$$

The step width Δt is determined adaptively by demanding that the changes in abundances per time step do not exceed a certain value.

We have adapted the code to the capabilities of modern computer technologies, which means that we have increased the internal numerical precision, implemented more precise integration routines and added several further numerical improvements. These corrections were necessary to remove numerical inaccuracies present in the available BBN codes.

---

[10]Some nuclear reactions can have both high forward and backward reaction rates. A naive solution of the differential equations for the nuclear reactions via $y(t_o + \Delta t) = y(t_0) + \Delta t \cdot y'(t_0)$ would not take into account that produced nuclei might be immediately destroyed after creation via the back reaction process.

6.8. OBSERVATIONAL SITUATION AND UNCERTAINTIES

In general, the uncertainties of the input parameters for BBN result in much higher uncertainties of the resulting abundances than do the numerical inaccuracies of the code. However, in the next chapter we will derive linearized dependences of the final abundances Y on input parameters X, $\partial \ln Y/\partial \ln X$. This will be achieved by slightly varying the input parameters X and observing the resulting changes in the abundances Y, where the variations in X are much below their uncertainty in order to get the correct linear order approximation. Hence, we have to ensure that the changes in the final results of the BBN simulations under small changes of the input parameters are not affected by numerical effects but solely by changes in the physical processes.

6.8 Observational situation and uncertainties

One of the biggest successes of standard BBN is the matching of theoretically predicted and observed primordial abundances for major elements. For a review of the theoretical and observational status and obstacles see for instance [Steigman05].

6.8.1 ^{4}He

The highest precision observation is that of the ^{4}He abundance (conventionally written Y_P). The post-BBN evolution of ^{4}He is simple. In stars, hydrogen is burned to ^{4}He which increases the abundance of ^{4}He above its primordial value. Hence one expects the ^{4}He abundance to decrease once one goes to stars with lower metallicity[11], and a ^{4}He "plateau" is expected for sufficiently low metallicity. Olive et al. [OliveSkillman04] analyze 8 systems with very low oxygen abundance, which are displayed in Fig. 6.3. They argue that the observational data indicated a primordial abundance of

$$Y_P = 0.249 \pm 0.009 , \qquad (6.57)$$

which we take to be a 1σ range. However, given the probable dominance of systematic effects, instead of using 2-σ bounds to later determine the range of allowed variations[12], we rather use the "conservative allowable range" of Y_P given in [OliveSkillman04] as

$$0.232 \leq Y_P \leq 0.258. \qquad (6.58)$$

6.8.2 Deuterium

For deuterium, the post-BBN evolution is as clear as it is for ^{4}He. Deuterium which was formed during primordial nucleosynthesis is burned in stars to ^{3}He and higher elements. Also, any deuterium which is newly produced in stars is immediately burned into ^{3}He as deuterium is the most weakly bound light nucleus. Hence, the deuterium abundance should increase when going back in time (opposed to ^{4}He, which decreases). However, the almost identical absorption spectra of D and "normal" hydrogen (^{1}H) complicate spectroscopic determinations of deuterium significantly.

---

[11]Elements with higher mass number than Helium (which are in astronomy called "metals") are only produced in stars. Hence a low metallicity of a system is a strong indication of element abundances which are assumed to be close to the primordial ones.

[12]See Sec. 9.1.

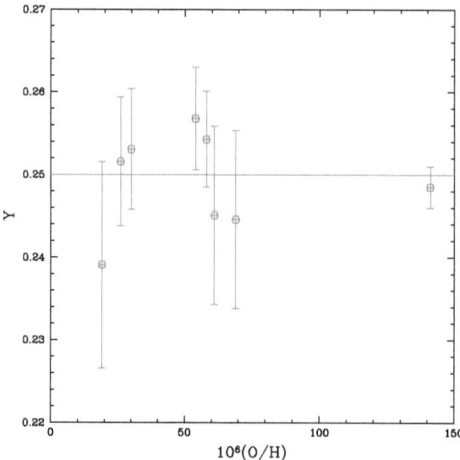

Figure 6.3: ^{4}He abundances versus oxygen abundance from [Steigman05]. The solid line is the weighted mean of for the 8 systems. Data from [OliveSkillman04]

Due to this complication, the determination of the primordial deuterium abundance follows from a small number of observed systems. One basically observes absorption spectra of interstellar deuterium from sightlines of Quasars which have to fulfill certain quality conditions (see [Kirkman03] for details). As D/H is of the order 10^{-5}, the hydrogen column density N_{HI} must be large enough in order to be able to observe deuterium with modern high-resolution spectrographs[13]. We use the determinations of [O'Meara06, Kirkman03] which give a value of

$$D/H = (2.8 \pm 0.4) \times 10^{-5}. \qquad (6.59)$$

The analysis of [O'Meara06] is displayed in Fig. 6.4, where the large scatter between values determined from different systems should be noted. After our studies have been performed, a new analysis appeared [Pettini08] which reduces the error of D/H by a factor of 2.

6.8.3 ^{3}He

The post-BBN development of ^{3}He is quite complex, as it is both produced and destroyed in stars. As quantitative analyses of these competing processes are quite model dependent, the primordial ^{3}He abundance cannot be extracted easily from spectra of old stars. This complexity is revealed by determinations of the ^{3}He abundance, which typically have a large scatter as can be seen in Fig. 6.5. Hence the common understanding is that ^{3}He abundance determinations cannot be considered as good tracers for the primordial abundance [Vangioni-Flam02].

---

[13]The column density N_{HI} gives a two-dimensional measure of the density (here number density) of a cloud of material, measured in cm^{-2}.

6.8. OBSERVATIONAL SITUATION AND UNCERTAINTIES

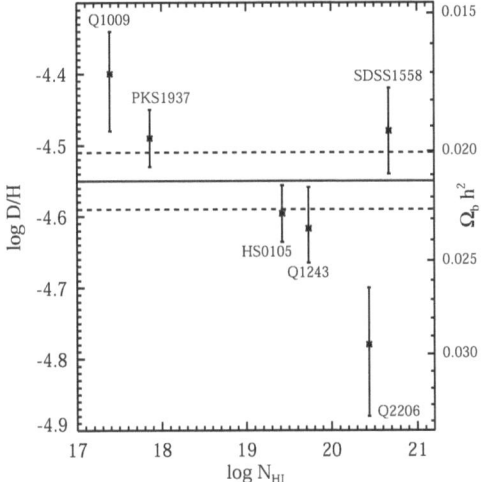

Figure 6.4: Deuterium versus neutral column density N_{HI} for a set of low-metallicity absorption spectra along QSO sightlines. From [O'Meara06]. (N_{HI} in cm^{-2}, its value is of no relevance here). The horizontal line represents the weighted mean, the right axis gives the derived SBBN values for $\Omega_b h^2$.

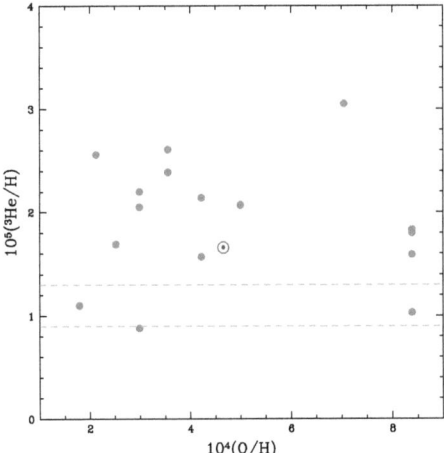

Figure 6.5: ^{3}He abundances versus oxygen from [Steigman05]. Data from [Bania02]. The dashed lines show the 1σ band given by [Bania02], the blue spot indicates the ^{3}He abundance for the pre-solar nebula.

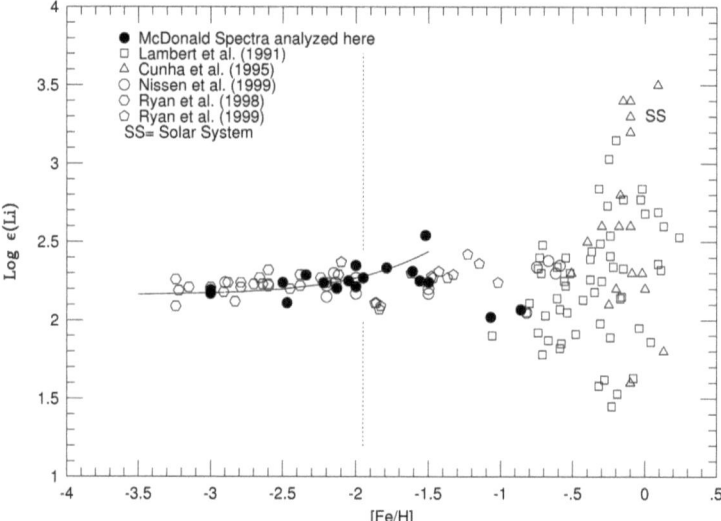

Figure 6.6: ^{7}Li abundances versus metallicity from [Steigman05]. $\log \epsilon(\text{Li}) \equiv 12 + \log(\text{Li/H})$

6.8.4 ^{7}Li

The post-BBN ^{7}Li abundance increases due to cosmic ray spallation/fusion reactions and ^{7}Li production in stars. Even though ^{7}Li is also easily destroyed inside stars, some stars observationally appear to be "super-lithium rich" [Steigman05], supporting the assumption that (most) stars are in fact net producers of lithium. Hence, one expects the lithium abundance to decrease when going back in time, which is supported by observations of the ^{7}Li abundance which show a plateau at old, metal-poor halo stars (see Fig. 6.6). One assumes that the plateau value is closely related to the primordial one.

The most recent determinations of the abundance are quite small: ^{7}Li/H $= (1.3 \pm 0.3) \cdot 10^{-10}$ [Bonifacio06] (see also [Asplund05]). It has been suggested that there are unresolved systematic errors relating to the effective temperature of the stars [Charbonnel05] which may imply a value as large as ^{7}Li/H $= (1.64 \pm 0.3) \cdot 10^{-10}$. To account for this possible systematic, we adopt a value

$$^7\text{Li/H} = (1.5 \pm 0.5) \times 10^{-10}. \tag{6.60}$$

It turns out that the observed ^{7}Li abundance is about a factor of 3 smaller than the standard theoretical prediction, which is called the "lithium problem".

6.8. OBSERVATIONAL SITUATION AND UNCERTAINTIES

Abundance	Observational	Theoretical
D/H	$(2.8 \pm 0.4) \times 10^{-5}$	$(2.61 \pm 0.04) \times 10^{-5}$
Y_P	0.249 ± 0.009	0.2478 ± 0.0002
^{7}Li/H	$(1.5 \pm 0.5) \times 10^{-10}$	$(4.5 \pm 0.4) \times 10^{-10}$

Table 6.2: Current observational and theoretical primordial abundances

6.8.5 ^{6}Li

A possible detection of ^{6}Li was discussed in [Asplund05], who claim to have found signals of ^{6}Li in nine stars at the $\geq 2\sigma$ significance level. Their observation suggests a ^{6}Li plateau at the level of

$$^6\text{Li/H} \approx 6.2 \times 10^{-12} \,. \tag{6.61}$$

If this detection is correct, the ^{6}Li abundance is about a factor 100 larger than the SBBN prediction. However, given the unclear observational status and post-BBN history of this isotope, we do not include ^{6}Li in the final analysis.

6.8.6   Theoretical predictions

Theoretically predicted primordial abundances also come with an error, mainly due to cross-section uncertainties. Our numerical procedures do not provide error estimates, so we adopt the 1σ ranges from [Serpico04], using a baryon density $\Omega_b h^2 = 0.0224$ [ManganoTalk07, HinshawWMAP5]:

$$\begin{aligned} \text{D/H} &= (2.61 \pm 0.04) \times 10^{-5} \\ ^3\text{He/H} &= (1.03 \pm 0.03) \times 10^{-5} \\ Y_P &= 0.2478 \pm 0.0002 \\ ^7\text{Li/H} &= (4.5 \pm 0.4) \times 10^{-10}. \end{aligned} \tag{6.62}$$

A compilation of the recent observational and theoretical primordial abundances which we will use for BBN is given in Tab. 6.2.

Chapter 7

BBN with varying constants

7.1 Nuclear and fundamental parameters

Big Bang Nucleosynthesis happens at energies in the keV regime, much below the energy scale of the Standard Model of particle physics where the parameters which are nowadays considered as fundamental (Tab. 2.1) are relevant. In particular, the theory of QCD cannot be applied at BBN as it happens well below the QCD scale $\Lambda_{QCD} \approx 200$ MeV. Quarks have combined into nucleons and the strong coupling constant α_S is in the non-perturbative regime. Hence, rather than the parameters of the SM, a set of effective low-energy parameters (masses of nucleons, binding energies, cross-sections and lifetimes) enters in the BBN simulations. We call these parameters "nuclear parameters" which stresses that those are the actual relevant (effective) parameters which enter in the nuclear physics processes.

7.2 Nuclear parameters relevant for BBN

The following set of effective low-energy ("nuclear") parameters enters the BBN simulations:

- Neutron and proton mass m_n, m_p, combined to
- Average nucleon mass $m_N := (m_n + m_p)/2$
- Neutron proton mass difference $Q_N := m_n - m_p$
- Neutron lifetime τ_n
- Binding energies of D, T, ^{3}He, ^{4}He, ^{6}Li, ^{7}Li, ^{7}Be.

Additionally, several fundamental parameters are also used in the BBN simulation:

- Electron mass m_e
- Gravitational constant G_N
- Fine structure constant α.

7.3. NUCLEAR PARAMETER DEPENDENCE

The baryon to photon ratio η enters as a further important parameter which is of cosmological origin. If one had a closed picture for the origin of our Universe, this ratio would derive from "fundamental" parameters, but lacking any theory this value only enters as an observed cosmological quantity. We combine all parameters which enter in the BBN simulation into a set X_i which we will call "set of nuclear parameters".[1]

7.3 Nuclear parameter dependence and the response matrix

We consider the set of primordial abundances Y_a with $a = $ (D, ^{3}He, ^{4}He, ^{6}Li, ^{7}Li) and study its dependence on the variation of our set of nuclear physics parameters X_i. Here the index i denotes the parameters which enter the calculation of nuclear abundances, listed in the preceding section. Our central quantity is the response matrix C with matrix elements [MSW04]

$$c_{ai} = \frac{\partial \ln Y_a}{\partial \ln X_i}. \tag{7.1}$$

It indicates the leading linear dependence for small deviations of the abundances about the values obtained given the nuclear parameters inferred from present laboratory experiments. The matrix C is extracted by varying the quantities X_i independently in the BBN code, a procedure which includes variation of the reaction cross-sections and rates that have a physical dependence on X_i. If all variations in parameters are taken to be small, all necessary information can indeed be extracted from the response matrix.

Our results for the nuclear response matrix are shown in Tab. 7.1. The first thirteen rows constitute the transposed nuclear response matrix C^T. We also quote the dependence of the abundances on η in the last row. Values are quoted to 2 d. p. or to 2 sig. fig. when the magnitude exceeds 1.

---

[1] Note that for instance variations with respect to the nuclear parameter α are different from variations with respect to the fundamental parameter α. This is due to the fact that variations w.r.t. the fundamental parameter incorporate *all* dependences, whereas variations w.r.t. the nuclear parameter only include *some* dependences. For instance the dependence of τ_n on α will only be included in the fundamental parameter dependence.

$\partial \ln Y_a / \partial \ln X_i$	D	^{3}He	^{4}He	^{6}Li	^{7}Li
τ_n	0.41	0.15	0.73	1.4	0.43
Q_N	0.83	0.31	1.55	2.9	1.00
m_N	3.5	0.11	-0.07	2.0	-12
B_D	-2.8	-2.1	0.68	-6.8	8.8
B_T	-0.22	-1.4	0	-0.20	-2.5
B_{3He}	-2.1	3.0	0	-3.1	-9.5
B_{4He}	-0.01	-0.57	0	-59	-57
B_{6Li}	0	0	0	69	0
B_{7Li}	0	0	0	0	-6.9
B_{7Be}	0	0	0	0	81
G_N	0.94	0.33	0.36	1.4	-0.72
α	2.3	0.79	0.00	4.6	-8.1
m_e	-0.16	-0.02	-0.71	-1.1	-0.82
η	-1.6	-0.57	0.04	-1.5	2.1

Table 7.1: Response matrix C, dependence of abundances on nuclear parameters.

Chapter 8

From nuclear to fundamental parameters

8.1 From nuclear to fundamental parameters

Looking at the Standard Model of particle physics, it is clear that the nuclear parameters given in the preceding section (Tab. 7.1) are degenerate in a sense that the set of 10 non-fundamental parameters only depends on about 5 to 6 fundamental parameters. Hence, in a next step, we will derive relations between a set of Standard Model parameters G_k and the nuclear physics parameters X_i. This is encoded in a second response matrix F with entries

$$f_{ik} = \frac{\partial \ln X_i}{\partial \ln G_k}. \tag{8.1}$$

The variation of abundances with respect to the fundamental parameters G_k is then expressed by the "fundamental response matrix" R with elements r_{ak},

$$\frac{\Delta Y_a}{Y_a} = r_{ak} \frac{\Delta G_k}{G_k}. \tag{8.2}$$

The matrix R is obtained from C and F by simple matrix multiplication,

$$R = CF. \tag{8.3}$$

We consider the following six fundamental parameters G_k:

- Gravitational constant G_N
- Fine structure constant α
- Electron mass m_e
- Light quark mass difference $\delta_q \equiv m_d - m_u$
- Averaged light quark mass $\hat{m} \equiv (m_d + m_u)/2 \propto m_\pi^2$
- Higgs v.e.v. $\langle \phi \rangle$.

An additional parameter is the strange quark mass m_s. We have omitted it from our treatment of BBN, as it enters the parameters (binding energies, neutron and proton mass) with high uncertainty. We have computed that the final dependence of abundances on the strange quark mass due to the known proton and neutron dependence is less than 3% of that of the light quark mass dependence (given in Tab. 8.2), hence the dependence on m_s is much lower than the model uncertainty[1] (*e.g.* for nuclear binding energies). However, in our study of variations in Chap. 11 we will include the strange quark mass contributions, as the parameters studied there are more directly influenced by strange effects. This is why we will also include the strange contribution in our following derivation of fundamental parameter dependences.

In the next sections, we give details on how the nuclear parameters depend on our set of fundamental parameters. In effective theories of nuclear forces, the pion appears as an effective mediator of the strong force. Hence, we will also introduce the pion mass m_π as an intermediate parameter.

8.1.1 Pion mass

The pion as the lightest of all mesons is the dominant mediator of the strong interaction in effective low energy theories of QCD. Its light mass is due to the fact that the pion is the pseudo Goldstone boson of the only slightly broken chiral symmetry of QCD. This fact implies that one can make several predictions about pion properties in chiral perturbation theory. In first order chiral perturbation theory, one obtains the famous Gell-Mann-Oakes-Renner relation [Gell-Mann68], which relates the pion mass to fundamental parameters[2][Gasser82],

$$m_\pi^2 \simeq -\frac{1}{f_\pi^2}(m_u + m_d)\langle 0|\bar{u}u + \bar{d}d|0\rangle \ . \tag{8.4}$$

The nonvanishing v.e.v. of $\bar{u}u$ and $\bar{d}d$ is a measure of the chiral asymmetry of the vacuum, signaling the spontaneous breakdown of chiral symmetry [Gasser82]. As can be seen from Eq. (8.4), the values of $\langle \bar{q}q \rangle$ are negative. f_π is the decay constant of the pion, which does not depend on quark masses at first order and which has a value of [PDG08]

$$f_\pi \approx 130 \text{ MeV} \ . \tag{8.5}$$

The formulas mentioned above are pure chiral perturbation theory equations neglecting any electromagnetic contributions. In fact, the pion mass difference of $m_{\pi^\pm} - m_{\pi^0} = 4.6$ MeV is dominantly due to electromagnetic effects [Gasser82]: up to a small uncertainty of 0.1 MeV the virtual photon cloud surrounding the $\pi^\pm$ accounts for the π^0-$\pi^\pm$ mass difference,

$$m_{\pi^\pm}^\gamma - m_{\pi^0}^\gamma = 4.6 \pm 0.1 \text{ MeV}$$

with $m_{\pi^0}^\gamma \approx 0$. We can safely neglect this contribution and only work with the first order dependence on the light quark mass, which can be read off from Eq. (8.4),

$$m_\pi^2 \propto \hat{m} \tag{8.6}$$

---

[1] However, it might be that binding energies get a substantial contribution from strange effects which we so far cannot quantify [Flambaum02].
[2] Note that there are in fact three pions, π^+, π^- and π^0 with roughly the same mass ($m_{\pi^\pm} = 139.6$ MeV, $m_{\pi^0} = 135.0$ MeV). We will only work with the first order dependences on fundamental parameters which are the same for all three pions. Further details on specific mass contributions can be found in [Gasser82].

8.1. FROM NUCLEAR TO FUNDAMENTAL PARAMETERS

or
$$\Delta \ln m_\pi = \frac{1}{2}\Delta \ln \hat{m} \ . \tag{8.7}$$

8.1.2 Neutron and proton mass

The different contributions to the neutron and proton mass and the origin of the proton neutron mass difference have been studied by Gasser and Leutwyler more than 25 years ago [Gasser75, Gasser82]. Using "improved chiral perturbation theory", they estimate the contributions coming from electromagnetic and (effective) strong self energy as well as from different quark masses. Knowing the properties of electron proton scattering they could estimate the electromagnetic contributions to the masses at Born level[3] from the electromagnetic form factors as

$$\begin{aligned} m_p^\gamma &= 0.63 \text{ MeV} \\ m_n^\gamma &= -0.13 \text{ MeV} \end{aligned} \tag{8.8}$$

to a very high accuracy. A further electromagnetic correction might come from possible intermediate states, a contribution which could not be calculated but has been estimated to

$$\Delta m_{res} = \pm 0.2 \text{ MeV} \ . \tag{8.9}$$

Thus, the total electromagnetic contributions to the nucleon masses are the values given in (8.8) with an error of ± 0.2 MeV. For the neutron-proton electromagnetic mass difference one obtains

$$(m_n - m_p)^\gamma = -0.76 \pm 0.30 \text{ MeV} \ . \tag{8.10}$$

The bare QCD masses of the nucleons are hence

$$\begin{aligned} m_p^{QCD} &= 937.64 \pm 0.20 \text{ MeV} \ , \\ m_n^{QCD} &= 939.70 \pm 0.20 \text{ MeV} \ . \end{aligned} \tag{8.11}$$

The difference in neutron and proton mass of

$$(m_n - m_p)^{QCD} = 2.05 \pm 0.30 \text{ MeV} \tag{8.12}$$

can be explained in terms of the different quark content, $p = (uud)$, $n = (udd)$, which turns out to be in lowest order quark mass expansion [Gasser82]

$$(m_n-m_p)^{QCD} = (m_d-m_u)\frac{1}{2m_N}<p|\bar{u}u-\bar{d}d|p> = \delta_q \frac{1}{2m_N}<p|\bar{u}u-\bar{d}d|p> \ . \tag{8.13}$$

Equations (8.10), (8.12) and (8.13) can be combined into

$$\Delta Q_N \simeq (-0.76\Delta \ln \alpha + 2.05 \Delta \ln \delta_q) \text{ MeV}$$

$$\Rightarrow \Delta \ln Q_N \simeq (-0.59\Delta \ln \alpha + 1.59\Delta \ln \delta_q) \ . \tag{8.14}$$

Subtracting the electromagnetic contributions, one is left with the pure QCD masses as given in Eqn (8.11), which is far from only being the sum of the masses of

---
[3]Born level contributions are proportional to α, so a rescaling of α by a certain factor scales the values given in Eq. (8.8) by the same factor.

the three valence quarks of the nucleons. A recent study using "heavy baryon chiral perturbation theory" [Borasoy96] yields that the baryon mass in the chiral limit is

$$\overset{\circ}{m} = 767 \pm 110 \text{ MeV}. \tag{8.15}$$

This quantity is purely due to QCD effects which only depend on Λ_{QCD}, and since we are holding this quantity fixed, $\overset{\circ}{m}$ remains constant. The contributions to the final nucleon mass come from the u, d, and s quarks, whose contributions can be estimated using the two quantities [Borasoy96]

$$\sigma_{\pi N}(0) := \hat{m} <p|\bar{u}u + \bar{d}d|p> = 45 \pm 10 \text{ MeV} \tag{8.16}$$

$$y := \frac{2 <p|\bar{s}s|p>}{<p|\bar{u}u + \bar{d}d|p>} = 0.21 \pm 0.20 . \tag{8.17}$$

They are connected to the quark mass dependence of the nucleons via the Feynmann-Hellmann theorem [Feynman39, Hellmann37],

$$\sigma_{\pi N}(0) = \hat{m} \frac{\partial m_N}{\partial \hat{m}} \tag{8.18}$$

and equivalently

$$<p|\bar{s}s|p> = \frac{\partial m_N}{\partial m_s} . \tag{8.19}$$

This can be translated into

$$\frac{\partial \ln m_N}{\partial \ln \hat{m}} = m_N^{-1} \sigma_{\pi N}(0) = 0.048 \pm 0.011 \tag{8.20}$$

$$\frac{\partial \ln m_N}{\partial \ln m_s} = \frac{m_s}{\hat{m}} \frac{y}{2} m_N^{-1} \sigma_{\pi N}(0) = 0.12 \pm 0.12 \tag{8.21}$$

using $\frac{m_s}{\hat{m}} \approx 25$ [PDG08].

8.1.3 Neutron lifetime

In the electroweak model the neutron lifetime can be evaluated analytically. One derives [WeinbergGRT] (see also Eq. (6.33) with zero temperature),

$$\tau_n^{-1} = \lambda_{n->p+e^-+\bar{\nu}_e} = \frac{1+3g_A^2}{2\pi^3} G_F^2 \int_{m_e}^{Q_N} x^2 (Q_N - x)^2 \sqrt{1 - \frac{m_e^2}{x^2}} dx . \tag{8.22}$$

Using $q := \frac{Q_N}{m_e}$, this integral evaluates to

$$\lambda_{n->p+e^-+\bar{\nu}_e} = \frac{1+3g_A^2}{120\pi^3} G_F^2 m_e^5 \left[\sqrt{q^2-1} (2q^4 - 9q^2 - 8) + 15q \ln \left(q + \sqrt{q^2-1} \right) \right] . \tag{8.23}$$

Here, G_F is the Fermi coupling constant, which is related to the Higgs v.e.v. via Eq. (4.21) and g_A is the nucleon axial vector coupling constant, $g_A \approx 1.27$ [PDG08]. Recent developments in lattice QCD, combined with chiral perturbation theory, allow to compute g_A, yielding [Hemmert03]

$$g_A = (1.2 \pm 0.1) - 6.9 \left(\frac{m_\pi}{1 \text{ GeV}} \right)^2 \ln \frac{m_\pi}{1 \text{ GeV}} - (10.4 \pm 4.8) m_\pi^2 \text{ GeV}^{-2} + \mathcal{O}(m_\pi^3) . \tag{8.24}$$

Obviously the chiral value $\overset{\circ}{g_A} = 1.2 \pm 0.1$ dominates the physical value of g_A, and for the derivative we obtain

$$\frac{\partial \ln \tau_n}{\partial \ln m_\pi} = 0.006 \implies \frac{\partial \ln \tau_n}{\partial \ln \hat{m}} = 0.003. \quad (8.25)$$

Thus g_A can be assumed constant in our treatment. Using the known dependences of Q_N from the preceding section, we arrive at a fundamental parameter dependence of the neutron lifetime of

$$\Delta \ln \tau_n = 3.86 \Delta \ln \alpha + 4 \Delta \ln \langle \phi \rangle + 1.52 \Delta \ln m_e - 10.4 \Delta \ln \delta_q. \quad (8.26)$$

8.1.4 Binding energies

The dependence of nuclear binding energies on the pion mass and α have been estimated in [Pudliner97] and [Pieper01] using quantum Monte Carlo calculations with realistic models of nuclear forces (similar values for the α dependence appear in [Nollett02]). We use the Pudliner and Pieper values which give for the dependence of binding energies on α:

$$\Delta \ln(B_D, B_T, B_{3He}, B_{4He}, B_{6Li}, B_{7Li}, B_{7Be}) =$$
$$(-0.0081, -0.0047, -0.093, -0.030, -0.054, -0.046, -0.088) \Delta \ln \alpha. \quad (8.27)$$

The pion mass determines the range of attractive nuclear forces, and the quantum Monte Carlo calculations of [Pudliner97, Pieper01] which include pion exchange accurately reproduce many experimental properties. One-pion exchange and two-pion exchange are dominant contributions within the expectation values of the two- and three-nucleon potentials respectively. Currently such studies have not been extended to determine the functional dependence of binding energies on the pion mass in general [Flambaum07, Damour07]. This dependence would in any case have uncertainties due to subleading effects of pion mass (or equivalently light quark masses) on other terms in the nucleon-nucleon potential [BeaneSavage02].

However, the dependence of the deuteron binding energy on the pion mass has been extensively studied within low-energy effective theory [Epelbaum02, BeaneSavage02]: the result may be expressed as

$$\Delta \ln B_D = r \Delta \ln m_\pi = \frac{r}{2} \Delta \ln \hat{m} \quad (8.28)$$

for small variations about the current value [YooScherrer02], with $-10 \leq r \leq -6$.[4] We will also take this dependence as a guide for the likely pion mass dependence of other binding energies. Although the size of the deuteron binding appears due to an accidental cancellation between attractive and repulsive forces, its derivative with respect to m_π (which is just B_D/m_π times r) is not expected to be subject to any cancellation. We also expect that the pion contribution to the total binding energy should increase with the number of nucleons; a proportionality to $(A-1)$ seems reasonable to obtain correct scaling at both small and large A^5. Hence to estimate the effect of pion mass on the binding energy of a nucleus B_i we set

$$\frac{\partial B_i}{\partial m_\pi} = f_i(A_i - 1)\frac{B_D}{m_\pi}r \simeq -0.13 f_i(A_i - 1), \quad (8.29)$$

---
[4]Our definition of r differs by a sign from [YooScherrer02].
[5]Flambaum et al. [Flambaum07] obtain roughly a scaling of type $(A-1)$ for nuclei with $3 \leq A \leq 7$, but their results have a large uncertainty.

taking $r \simeq -8$. The numerical constants f_i are expected to be of order unity, but will differ between light nuclei due to peculiarities of the shell structure, *etc.* Our normalization corresponds to $f_D = 1$. Then the nontrivial dependences of nuclear parameters on $\hat{m}$ are

$$\Delta \ln(B_{\rm D}, B_{\rm T}, B_{\rm 3He}, B_{\rm 4He}, B_{\rm 6Li}, B_{\rm 7Li}, B_{\rm 7Be}, m_N) \simeq$$
$$(0.5r, 0.26f_{\rm T}r, 0.29f_{\rm 3He}r, 0.12f_{\rm 4He}r, 0.17f_{\rm 6Li}r, 0.17f_{\rm 7Li}r, 0.18f_{\rm 7Be}r, 0.048)\Delta \ln \hat{m} \ , \tag{8.30}$$

where the dependence of m_N is taken from Eq. (8.20). For the $\hat{m}$ dependence of abundances due to the variation of binding energies we then have

$$\left.\frac{\partial \ln Y_a}{\partial \ln \hat{m}}\right|_B = \frac{r}{2}\sum_i f_i \frac{(A_i - 1)B_D}{B_i}\frac{\partial \ln Y_a}{\partial \ln B_i} \ , \tag{8.31}$$

where the dependence $\frac{\partial \ln Y_a}{\partial \ln B_i}$ is obtained from the BBN code by varying the binding energies B_i. Taking account also of the small effect of $\hat{m}$ on the nucleon mass m_N, the resulting dependence of abundances on $\hat{m}$ is

$$\frac{\partial \ln Y_{\rm D}}{\partial \ln \hat{m}} \simeq 11 + 0.5f_T + 5f_{\rm 3He}$$
$$\frac{\partial \ln Y_{\rm 3He}}{\partial \ln \hat{m}} \simeq 8 + 3f_T - 7f_{\rm 3He}$$
$$\frac{\partial \ln Y_{\rm 4He}}{\partial \ln \hat{m}} \simeq -2.7$$
$$\frac{\partial \ln Y_{\rm 6Li}}{\partial \ln \hat{m}} \simeq 27 + 0.4f_{\rm T} + 7f_{\rm 3He} + 55f_{\rm 4He} - 96f_{\rm 6Li}$$
$$\frac{\partial \ln Y_{\rm 7Li}}{\partial \ln \hat{m}} \simeq -36 + 5f_T + 22f_{\rm 3He} + 54f_{\rm 4He} + 9f_{\rm 7Li} - 115f_{\rm 7Be} \ , \tag{8.32}$$

where we have neglected subleading terms. Even if we consider that some contributions could cancel against one another due to the values of the f_i, the magnitude of these variations is striking, particularly concerning the lithium abundances. To get an idea of the possible effect of cancellations, we may set all f_i to unity and find the dependences

$$\Delta \ln(Y_{\rm D}, Y_{\rm 3He}, Y_{\rm 4He}, Y_{\rm 6Li}, Y_{\rm 7Li}) \simeq (17, 5, -2.7, -6, -61)\Delta \ln \hat{m}. \tag{8.33}$$

One may also consider to what extent varying $\hat{m}$ or the pion mass may affect reaction cross-sections beyond the $npd\gamma$ reaction. It seems very likely that matrix elements would acquire nontrivial dependence on m_π; however, since the dependence of abundances on reaction cross-sections is relatively mild (see Table 6.1), the dependence via reaction matrix elements is unlikely to compete with the very large effects arising through the variation of binding energies.

8.2. THE RESPONSE MATRICES

$\partial \ln X_i/\partial \ln G_k$	G_N	α	$\langle\phi\rangle$	m_e	δ_q	$\hat{m}$
G_N	1	0	0	0	0	0
α	0	1	0	0	0	0
τ_n	0	3.86	4	1.52	-10.4	0
m_e	0	0	0	1	0	0
Q_N	0	-0.59	0	0	1.59	0
m_N	0	0	0	0	0	0.048
B_D	0	-0.0081	0	0	0	-4
B_T	0	-0.0047	0	0	0	$-2.1 f_T$
B_{3He}	0	-0.093	0	0	0	$-2.3 f_{3He}$
B_{4He}	0	-0.030	0	0	0	$-0.94 f_{4He}$
B_{6Li}	0	-0.054	0	0	0	$-1.4 f_{6Li}$
B_{7Li}	0	-0.046	0	0	0	$-1.4 f_{7Li}$
B_{7Be}	0	-0.088	0	0	0	$-1.4 f_{7Be}$

Table 8.1: Response matrix F, dependence of nuclear parameters X_i on fundamental parameters G_k

$\partial \ln Y_a/\partial \ln G_k$	D	^{3}He	^{4}He	^{6}Li	^{7}Li
G_N	0.94	0.33	0.36	1.4	-0.72
α	3.6	0.95	1.9	6.6	-11
$\langle\phi\rangle$	1.6	0.60	2.9	5.5	1.7
m_e	0.46	0.21	0.40	0.97	-0.17
δ_q	-2.9	-1.1	-5.1	-9.7	-2.9
$\hat{m}$	17	5.0	-2.7	-6	-61
η	-1.6	-0.57	0.04	-1.5	2.1

Table 8.2: Response matrix R, dependence of abundances Y_i on fundamental parameters G_k. All f_i are set to unity.

8.2 The response matrices

The results of the previous section can be summarized in the response matrix F defined in Eq. (8.1) which relates variations in fundamental parameters G_k to variations in nuclear parameters X_i. Its values are shown in Tab. 8.1. These results can be combined with the nuclear response matrix, Tab. 7.1, to the "fundamental response matrix" R defined in Eq. (8.2) according to Eq. (8.3). The matrix R relates variations in fundamental parameters to variations in final abundances and is shown in Tab. 8.2. This table is the central result of this part of the thesis. In treating the $\hat{m}$-dependences, which arise from the nuclear binding energies with their uncertain values of f_i, we have given the values which arise when setting all f_i to unity. In our further treatment of BBN we will neglect the model uncertainty[6] in the binding energies and always assume $f_i \equiv 1$.

---

[6] Given the current status of low-energy QCD, it seems hard to quantify possible ranges of uncertainty for the f_i.

$\partial \ln Y_a/\partial \ln G_k$	G_N	α	$\langle\phi\rangle$	m_e	δ_q	$\hat{m}$
^{4}He	0.41	1.94	3.36	0.389	-5.358	-1.59

Table 8.3: Dependence of abundances Y_i on fundamental parameters G_k found by [MSW04]

$\partial \ln Y_a/\partial \ln G_k$	D	^{3}He	^{4}He	^{6}Li	^{7}Li
G_N	0.4	0.6	0.3	0.3	-1.0
α	2.3	1.0	2.3	7.4	-9.5
$\langle\phi\rangle$	0.07	0.1	3.1	4.1	≈ 0.5

Table 8.4: Dependence of abundances Y_i on fundamental parameters from [Chamoun05] and [Landau04]

8.3 Comparison to other studies

The ^{4}He dependence was previously calculated in [MSW04] by semi-analytic methods [Esmailzadeh91]. Their findings are displayed in Tab. 8.3, our results for the dependence on fundamental parameters are similar. The dependence on G_N can be compared with the results of [Scherrer03] and [Chamoun05, Landau04], once one translates from units where G_N is constant to ours where Λ_{QCD} is constant. The latter ones also give values for the dependence on $\langle\phi\rangle$ (variation of $G_F \propto \langle\phi\rangle^{-2}$) and on α. Their results are shown in Tab. 8.4. They roughly match with our estimates most times except for a few numbers, e.g. the much smaller variations of D and ^{3}He under variations of the Higgs v.e.v. We assume that these deviations are due to the applied semi-analytical method of [Esmailzadeh91] which might not give appropriate results in the case of abundances other than ^{4}He. Bergström and Nollett [Bergstrom99, Nollett02] use the full BBN code [Kawano92] to constrain the variation of α, incorporating α-dependent reaction cross-sections as described in Sec. 6.6.3. From their graphical results we extract the dependences given in Tab. 8.5 which again match with our findings.

$\partial \ln Y_a/\partial \ln G_k$	D	^{3}He	^{4}He	^{6}Li	^{7}Li
α	≈ 3.5	≈ 0.9	≈ 2.0	-	≈ -7

Table 8.5: Dependence of abundances Y_i on α from [Bergstrom99] and [Nollett02]

Chapter 9

Constraints on variations of fundamental parameters

9.1 Bounds on separate variations of fundamental couplings

The first application of our results from the preceding chapter is in setting bounds on the variation of each fundamental parameter considered separately, under the assumption that only one parameter varies at once. We may consider three observational determinations of primordial abundances (see Sec. 6.8): deuterium, ^{4}He and ^{7}Li. However, the observed ^{7}Li abundance deviates by a factor two to three from the value predicted by standard BBN theory (SBBN), and systematic uncertainties related to stellar evolution exist [Korn06]. Thus, we use the former two, D and ^{4}He, to constrain the allowed variations of the fundamental constants individually. For deuterium we take 2σ limits; for ^{4}He we consider instead the "conservative allowable range" of [OliveSkillman04]. The resulting constraints are given in Tab. 9.1.

9.2 Variations of abundances in unified models

A major problem of using BBN to constrain variations of fundamental parameters is degeneracy. There are only three observational abundances (D, ^{4}He and ^{7}Li) but six fundamental parameters which can vary. Hence, the three observed values cannot be used to constrain a set of six parameters. In Chapter 4 we have introduced unified

-19%	$\leq \Delta \ln G_N \leq$	$+10\%$
-3.6%	$\leq \Delta \ln \alpha \leq$	$+1.9\%$
-2.3%	$\leq \Delta \ln \langle \phi \rangle \leq$	$+1.2\%$
-17%	$\leq \Delta \ln m_e \leq$	$+9.0\%$
-0.7%	$\leq \Delta \ln \delta_q \leq$	$+1.3\%$
-1.3%	$\leq \Delta \ln \hat{m} \leq$	$+1.7\%$

Table 9.1: Allowed individual variations (2σ or "conservative allowable range", see Sec. 6.8) of fundamental couplings.

models where the variations of fundamental couplings satisfy relations that reduce the number of free parameters. Here we apply GUT models in order to be able to visualize our findings for BBN. In the simplest case every variation of a parameter G_k is determined by a single underlying degree of freedom, for instance a variation of the unified coupling $\Delta \ln \alpha_X$. We may also eliminate $\Delta \ln \alpha_X$ in favor of some observable parameter, which we will choose to be the fine structure constant α.

For grand unified theories, it makes sense to change from a system of units with constant Λ_{QCD} to a system where M_{GUT}, the mass scale of the GUT, is set to constant. Furthermore, for the scenarios we will use in this section, we will take the Planck mass fixed relative to the unification scale, thus $\Delta(M_P/M_{GUT}) = 0$. We set all f_i to unity and, for simplicity, we take the Yukawa couplings to be constant[1], thus the electron and quark masses are proportional to $\langle \phi \rangle$,

$$\Delta \ln \frac{m_e}{M_{GUT}} = \Delta \ln \frac{\delta_q}{M_{GUT}} = \Delta \ln \frac{\hat{m}}{M_{GUT}} = \Delta \ln \frac{\langle \phi \rangle}{M_{GUT}}. \tag{9.1}$$

Finally we define an exponent γ which relates the variation of $\langle \phi \rangle$ with respect to M_{GUT} to the variation of Λ_{QCD}/M_{GUT} as

$$\frac{\langle \phi \rangle}{M_{GUT}} = \text{const.} \left(\frac{\Lambda_{QCD}}{M_{GUT}} \right)^\gamma. \tag{9.2}$$

We study three non-supersymmetric GUT scenarios, $\gamma = 0$, $\gamma = 1$ and $\gamma = 1.5$, and use the fine structure constant α to parametrize the variations. As has been shown in Sec. 4.6, the variation of α can be related to variations of particle masses, the GUT coupling α_X and the QCD invariant scale Λ_{QCD}, see equation (4.23).

Scenario $\gamma = 0$

In the first scenario, $\gamma = 0$, the Higgs v.e.v. and elementary fermion masses are all proportional to the unification scale, $\Delta(M_{GUT}/M_P, \langle \phi \rangle/M_P, m_{e,q}/M_P) = 0$. Then the variations of fundamental couplings are related as

$$\Delta \ln \frac{\Lambda_{QCD}}{M_{GUT}} = \Delta \ln \frac{\Lambda_{QCD}}{\langle \phi \rangle} = \frac{3\pi}{40\alpha} \Delta \ln \alpha \simeq 32.3 \Delta \ln \alpha. \tag{9.3}$$

In a system with constant Λ_{QCD} we get

$$\Delta \ln(G_N, \alpha, \langle \phi \rangle, m_e, \delta_q, \hat{m}) \simeq (64.5, 1, -32.3, -32.3, -32.3, -32.3) \Delta \ln \alpha. \tag{9.4}$$

We then obtain variations of abundances

$$\Delta \ln(Y_D, Y_{3He}, Y_{4He}, Y_{6Li}, Y_{7Li}) \simeq (-450, -130, 170, 380, 1960) \Delta \ln \alpha. \tag{9.5}$$

Scenario $\gamma = 1$

In the second scenario with $\gamma = 1$, all low-energy mass scales of particle physics are proportional to Λ_{QCD}. Setting Λ_{QCD} constant, we have

$$\Delta \ln(G_N, \alpha, \langle \phi \rangle, m_e, \delta_q, \hat{m}) \simeq (78, 1, 0, 0, 0, 0) \Delta \ln \alpha. \tag{9.6}$$

In this case the variations of abundances are not subject to the theoretical uncertainty of varying m_s/Λ_{QCD}. We obtain

$$\Delta \ln(Y_D, Y_{3He}, Y_{4He}, Y_{6Li}, Y_{7Li}) \simeq (77, 27, 30, 120, -68) \Delta \ln \alpha. \tag{9.7}$$

---
[1] See also the comments on variations of the Yukawa couplings in Sec. 11.1.

9.2. VARIATIONS OF ABUNDANCES IN UNIFIED MODELS

Scenario $\gamma = 1.5$

In the third unified scenario with $\gamma = 1.5$ we consider the case when the Higgs v.e.v. and fermion masses vary *more* rapidly (with respect to the unification scale) than the QCD scale Λ_{QCD} does. Converting back to a system where Λ_{QCD} is constant, we find that the variations of fundamental couplings are related as

$$\Delta \ln(G_N, \alpha, \langle\phi\rangle, m_e, \delta_q, \hat{m}) \simeq (87, 1, 21.5, 21.5, 21.5, 21.5)\Delta \ln \alpha. \quad (9.8)$$

The variations of abundances are then

$$\Delta \ln(Y_D, Y_{3He}, Y_{4He}, Y_{6Li}, Y_{7Li}) \simeq (430, 130, -65, -60, -1420)\Delta \ln \alpha. \quad (9.9)$$

9.2.1 Linear results

In Fig. 9.1 we show the abundance variations given by the three GUT models, as a function of the variation of α. First we plot only the linear dependence of abundances on α as given by the fundamental response matrix R (Tab. 8.2). We also show the 1σ observational bounds as highlighted regions. Also included in the plot is the effect on the standard BBN predictions of varying the baryon-to-photon ratio η over the 2σ range allowed by WMAP 3 year data[2], $5.7 \leq 10^{10}\eta \leq 6.5$.

It can be seen that in the $\gamma = 0$ scenario a reduction of α by about 0.025% (*i.e.* a fractional variation of -2.5×10^{-4}) would bring theory and observation into agreement within 2σ bounds, while remaining in the linear regime. Conversely, in the $\gamma = 1.5$ model an increase of α by about 0.04%, *i.e.* $\Delta \ln \alpha = 4 \times 10^{-4}$, brings theory and observation into agreement within 1σ bounds. Considering the variations of fundamental parameters in the three scenarios Eqns. (9.4), (9.6) and (9.8), the behavior of the weak scale $\langle\phi\rangle$ and fermion masses is decisive for the variation of abundances. However, if $\Delta \ln Y_a$ becomes larger than 1 (as in the case of ^{7}Li) the results are affected by higher order terms and the linear approximation can no longer be applied.

9.2.2 Nonlinear results

The linear analysis in the previous section suggests that it is possible, and may even be natural, to obtain a large negative variation in the ^{7}Li abundance, and considerably smaller variations in other measurable abundances: positive in the case of deuterium and negative for ^{4}He. Agreement between theory and data in all three abundances could then be possible for a narrow range of values in the variation of fundamental parameters. The required fractional variation in ^{7}Li is so large (a factor two or more in Y_{7Li}) that a linear analysis using matrix multiplication is inaccurate.

We improve the analysis by including the relations between nuclear and fundamental parameters and the three GUT relations for the fundamental parameters in our BBN simulation code. Then we run the code for a set of parameters $\Delta \ln \alpha$ and obtain the full nonlinear parameter dependence of abundances on variations of fundamental parameters. Note that this method is impractical to investigating the full parameter space: fundamental parameters span a six-dimensional and nuclear parameters a 13-dimensional parameter space which cannot be analyzed numerically due to

---

[2]Updating to WMAP5 values does not lead to any significant change, there the 2σ range is $5.9 \leq 10^{10}\eta \leq 6.5$.

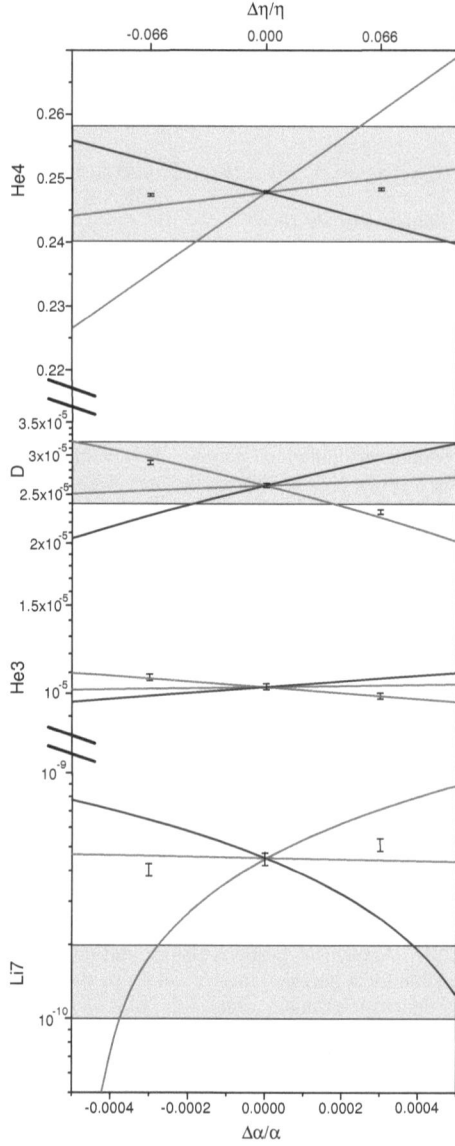

Figure 9.1: Variation of primordial abundances with α in three GUT scenarios. Red lines show the first $\gamma = 0$ scenario; green lines the second $\gamma = 1$ and blue lines the third $\gamma = 1.5$ scenario. Highlighted regions give the observational 1σ limits (as explained in Sec. 6.8.3, no observational limits can be given for ^{3}He). Error bars indicate the standard BBN abundances with theoretical 1σ error [Serpico04], for three different values of η about the WMAP central value, as indicated on the upper horizontal axis.

9.2. VARIATIONS OF ABUNDANCES IN UNIFIED MODELS

time reasons. It is only practicable if the dimensionality of the parameter space is reduced by applying unification relations.

For most nuclear parameters X_i the dependence on fundamental parameters G_k is only known to linear order or the nonlinear dependence involves a high uncertainty. However, we tested that for the unified models considered here, the fractional variations in the nuclear parameters X_i remain small, well below 0.1. A linear approximation for the relation between nuclear and fundamental parameters is therefore appropriate. The main nonlinear effects enter at the level of nuclear reactions, in equations where nuclear parameters as binding energies enter in a known power law dependence.

The nuclear parameters affecting most the large variation in ^{7}Li abundance are mainly the deuterium and ^{7}Be binding energies, with the ^{3}He and ^{4}He binding energies playing a smaller role. A decrease of B_D causes BBN to happen later, which means that the nucleon density is lower and reaction rates smaller. The abundances of $A > 4$ elements are rate-limited and thus decrease with decreasing B_D. This accounts for about two-thirds of the change in Y_{7Li}. In addition, the cross-section of the ^{3}He$(\alpha,\gamma)^7$Be reaction depends strongly on the Q-value, hence on the ^{7}Be binding energy. Both these effects are computationally under control, therefore we believe that the specific nonlinear dependence in the scenarios we consider is well estimated within our code. We show in Fig. 9.2 the primordial abundances including nonlinear effects, i.e. without using a linear approximation for the relation between Y_a and X_i. For our three GUT models we find a slightly different behavior of the ^{7}Li abundance, which now has an approximately power-law dependence on variation of α. It is only slightly more difficult to bring the present observational abundances into agreement with standard BBN and the WMAP determination of η ; still, if we allow a variation of $0.00045 \lesssim \Delta \ln \alpha \lesssim 0.0005$ in the $\gamma = 1.5$ model, the predicted abundances are all very close to the 1σ allowed regions.

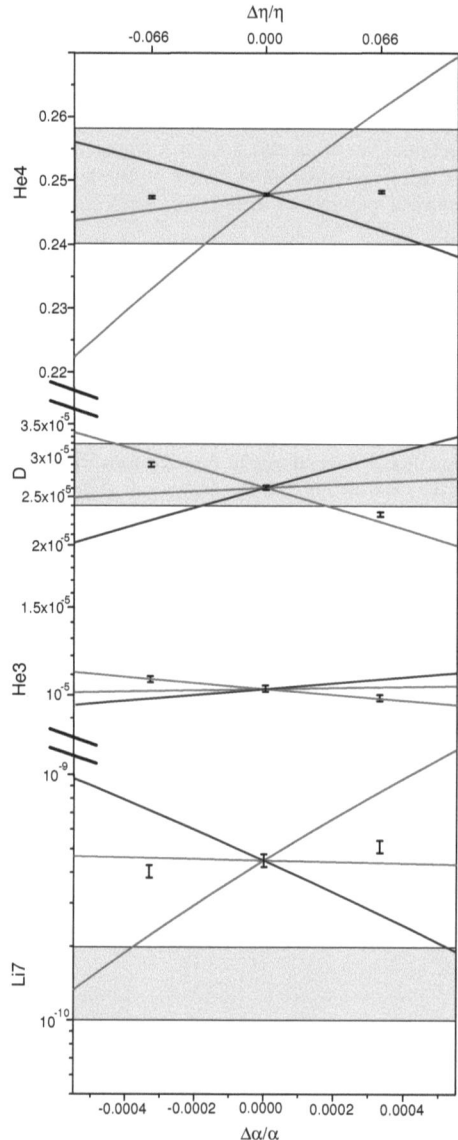

Figure 9.2: Variation of primordial abundances with α in three GUT scenarios including nonlinear effects. Labels as in Fig. 9.1.

Part III

Unifying cosmological and late-time variations

Chapter 10

Experimental tests of variations

In recent years possible variations in the constants of nature have been tested in various ways. Whilst direct laboratory measurements do not point towards any variation, some astrophysical tests yield slight variations. Still, those non-zero results neither constitute compelling evidence for variations since the applied fitting and analysis methods are still under debate, nor do the different non-zero results seem to be consistent with each other[1]. Hence the question if there are and have been variations remains open from the experimental side.

Here we review and discuss the observational data that we will consider in our effort to obtain a unified picture of time variation of couplings. We summarize the results that are most relevant for our analysis in Tab. 10.1.

10.1 BBN

The earliest processes for which Standard Model physics can be tested is BBN ($z \sim 10^{10}$). The influence of varying constants on BBN has been laid out in chapters 7 to 9. In an extension of our previous treatment, we include in Sec. 10.2.1 the possible effect of varying constants at last scattering (formation of the CMB) on the input parameter η of our BBN procedure.

The uncertainty in the η determination, $\eta = (6.20 \pm 0.16) \times 10^{-10}$ (WMAP5 plus BAO and SN, [WMAP5]) yields a further correlated error for the abundances, which can be treated using the method of [Fiorentini98]. For any given set of fundamental variations we can define

$$\chi^2 \equiv \sum_{i,j}(Y_i - Y_i^{obs})w_{ij}(Y_j - Y_j^{obs}), \qquad (10.1)$$

with the inverse weight matrix

$$w_{ij} = \left[\sigma_{ij}^{2,\eta} + \delta_{ij}(\sigma_{obs,i}^2 + \sigma_{th,i}^2)\right]^{-1}, \qquad (10.2)$$

---

[1] For a thorough comparison of measurements of variations, a more profound theoretical background is needed, as we will lay out in the rest of this thesis.

where
$$\sigma_{ij}^{2,\eta} \equiv Y_i Y_j \frac{\partial \ln Y_i}{\partial \ln \eta} \frac{\partial \ln Y_j}{\partial \ln \eta} \left(\frac{\Delta \eta}{\eta}\right)^2. \tag{10.3}$$

We take, as in Tab. 7.1,
$$\frac{\partial \ln(D/H, Y_p, {}^7Li/H)}{\partial \ln \eta} = (-1.6,\ 0.04,\ 2.1). \tag{10.4}$$

The $1(2)\sigma$ error contour is given by $\chi^2/\nu \leq 1(4)$ where ν is the number of degrees of freedom. As the final abundances depend on variations of all fundamental constants, we have to evaluate the variations allowed by BBN for every model separately.

In the light of complex astrophysics which may affect the extraction of the primordial ^{7}Li fraction, we also consider bounding the variations using deuterium and ^{4}He alone (see also Sec. 9.1). This yields a value consistent with zero for variations at BBN, since these abundances are consistent with standard BBN.

10.2   CMB

A further far-reaching test of varying constants is the cosmic microwave background (CMB) ($z \sim 10^3$). In principle, α and G_N are bounded by CMB observations. A variation of α affects the formation of the CMB through Thomson scattering and the recombination history. However, bounds on variations from CMB are typically very weak as there are significant degeneracies with other cosmological parameters [Martins03, Rocha03], see also the discussion in Sec. 10.2.1. Current bounds on α are [Rocha03]

$$0.95 < \frac{\alpha_{\text{CMB}}}{\alpha_0} < 1.02 \qquad (2\sigma). \tag{10.5}$$

The CMB anisotropies may also be used to constrain the variation of Newton's constant G_N. The resulting bound depends on the form of the variation of G_N from the time of CMB to now. Using a step function one finds [Chan07, Zahn02]

$$0.95 \leq \frac{G_N}{G_{N,0}} \leq 1.05 \qquad (2\sigma), \tag{10.6}$$

where the instantaneous change in G_N may happen at any time between now and CMB decoupling. Using instead a linear function of the scale factor a, the bound is

$$0.89 \leq \frac{G_N}{G_{N,0}} \leq 1.13 \qquad (2\sigma). \tag{10.7}$$

Note that here, as in most studies of time-dependent G_N, units are implicitly defined such that the elementary particle masses (and thus the mass of gravitating bodies, if gravitational self-energy is neglected) are constant. The relevant bound on *dimensionless* parameters concerns $G_N m_N^2 \equiv (m_N/M_P)^2 (8\pi)^{-1}$.

10.2.1 Effect of "varying constants" at CMB and η

In our study on BBN we used the WMAP determination of the baryon number density parameter $\eta \equiv n_B/n_\gamma$ directly to reduce by one the number of unknown parameters. However, we should also consider the effect of possible variations of fundamental parameters at the epoch of CMB decoupling, as CMB measurements are used to derive the value of η. Hence, it seems important to study how possible variations at the epoch of CMB decoupling affect the determination of η and hence also the outcomes of BBN simulations.

Fundamental parameters affecting the CMB are the proton and electron masses, the gravitational constant and the fine structure constant, as well as the mass of any dark matter particle present. In Planck units, these reduce to the particle masses and α. The relevant cosmological parameters are the amplitude, spectral index (and possible running, *etc.*) of primordial perturbations; the baryon, dark matter, dark energy (cosmological constant, *etc.*) and curvature densities normalized to the critical density; the Hubble constant; and the reionization optical depth. Of these, the baryon density $\Omega_b h^2$ will vary linearly with the proton mass in Planck units, for a fixed baryon-to-photon ratio η. Conversely, given a measurement of $\Omega_b h^2$, the correct value of η varies inversely with the proton mass. The conversion factor between $\Omega_b h^2$ and $\eta_{10} \equiv 10^{10}\eta$ is then (see Eq. (6.14))

$$273.9(m_p\sqrt{G_N})|_0 (m_p\sqrt{G_N})^{-1} \simeq 273.9(1 - \Delta\ln(m_N/M_P)_{|\text{CMB}}), \quad (10.8)$$

where we approximate the proton and neutron masses by their average m_N.

If, therefore, we allow the proton mass (or the gravitational constant, in QCD units) to vary arbitrarily at the CMB epoch, η is undetermined by WMAP and we must consider it as an extra free parameter or try to impose independent cosmological bounds. However, we impose that the size of variations away from the present value of m_p/M_P is a monotonically decreasing function of time: thus $\Delta\ln(m_N/M_P)_{|\text{CMB}} \leq \Delta\ln(m_N/M_P)_{|\text{BBN}}$. Hence we would have a self-consistent treatment of this parameter if the secondary discrepancies in primordial abundances due to an incorrectly estimated η were smaller than the primary effect of varying m_N/M_P at BBN. The maximum effect due to rescaling of η would occur when $\Delta\ln(m_N/M_P)_{|\text{CMB}} = \Delta\ln(m_N/M_P)_{|\text{BBN}}$, adding the nucleon mass induced variation in η,

$$\Delta\ln\eta = \Delta\ln(m_N/M_P)_{|\text{BBN}}. \quad (10.9)$$

In our treatment of BBN in the next chapter, we will consider this effect by studying two limiting cases. First, when $\Delta(m_N/M_P)_{|\text{CMB}} \ll \Delta(m_N/M_P)_{|\text{BBN}}$, then our previous results hold. In the second case, with $\Delta(m_N/M_P)_{|\text{CMB}} \simeq \Delta(m_N/M_P)_{|\text{BBN}}$, the value of η may be significantly rescaled.

10.3 Quasar absorption spectra

The observation of absorption spectra of distant interstellar clouds allows to probe atomic physics over large time scales. Comparing observed spectra with the spectra observed in the laboratory, together with the in general well known dependence of the spectra on fundamental constants, gives bounds on the possible variation of couplings. Different kinds of spectra (atomic, molecular, ...) are sensitive to different parameters, which we will list in the following paragraphs.

10.3. QUASAR ABSORPTION SPECTRA

Tests of α

Atomic spectra are primarily sensitive to α. Several groups using various methods of modeling and numerical analysis have published results; we quote here only the latest bounds. Murphy and collaborators [Murphy03.2] studied the spectra of 143 quasar absorption systems over the redshift range $0.2 < z_{abs} < 4.2$. Their most robust estimate is a weighted mean

$$\frac{\Delta \alpha}{\alpha} = (-0.57 \pm 0.11) \times 10^{-5}. \tag{10.10}$$

Dividing the data into low ($z < 1.8$) and high ($z > 1.8$) redshift subsamples, they obtain

$$z < 1.8, \quad N_{sys} = 77, \quad \langle z_{abs} \rangle = 1.07, \quad \frac{\Delta \alpha}{\alpha} = (-0.54 \pm 0.12) \times 10^{-5}$$

$$z > 1.8, \quad N_{sys} = 66, \quad \langle z_{abs} \rangle = 2.55, \quad \frac{\Delta \alpha}{\alpha} = (-0.74 \pm 0.17) \times 10^{-5}, \tag{10.11}$$

where N_{sys} is the number of absorption systems in the sample and $\langle z_{abs} \rangle$ is the averaged sample redshift.

In discussing unified models in Sec. 11.3, we will define various "epochs" for the purpose of collating data and comparing them with models over certain ranges of redshift. The 143 data points are then assigned to different epochs: we choose to put boundaries at $z = 0.81$ and $z = 2.4$, thus we obtain three sub-samples

$$z < 0.81, \quad N_{sys} = 18, \quad \langle z \rangle = 0.65, \quad \frac{\Delta \alpha}{\alpha} = (-0.29 \pm 0.31) \times 10^{-5}$$

$$0.81 < z < 2.4 \quad N_{sys} = 85, \quad \langle z \rangle = 1.47, \quad \frac{\Delta \alpha}{\alpha} = (-0.58 \pm 0.13) \times 10^{-5}$$

$$z > 2.4, \quad N_{sys} = 40, \quad \langle z \rangle = 2.84, \quad \frac{\Delta \alpha}{\alpha} = (-0.87 \pm 0.37) \times 10^{-5}. \tag{10.12}$$

Here we have used the "fiducial sample" of [Murphy03.1], the weighted average has been taken, and we have included [Murphyprivate] the 15 additional samples used in [Murphy03.2]. For convenience we will refer to these results as "Mα".

Further results have been obtained by Levshakov et al. [Levshakov07.1], and reported in [Fujii07]:

$$\frac{\Delta \alpha}{\alpha} = (-0.01 \pm 0.18) \times 10^{-5}, \quad z_{abs} = 1.15$$

$$\frac{\Delta \alpha}{\alpha} = (0.57 \pm 0.27) \times 10^{-5}, \quad z_{abs} = 1.84. \tag{10.13}$$

We note that the value for $z = 1.84$ has an opposite sign of variation to the Mα result, though the variation does not have high statistical significance. The observational situation is clearly unsatisfactory.

Tests of μ

Vibro-rotational transitions of molecular hydrogen H_2 are sensitive to $\mu \equiv m_p/m_e$. From H_2 lines of two quasar absorption systems (at $z = 2.59$ and $z = 3.02$) a variation is found [Reinhold06] of

$$\frac{\Delta \mu}{\mu} = (2.4 \pm 0.6) \times 10^{-5}, \tag{10.14}$$

taking a weighted average. We will refer to this result as "Rμ" after Reinhold et al. The individual systems yield [Reinhold06]

$$\frac{\Delta\mu}{\mu} = (2.78 \pm 0.88) \times 10^{-5}, \qquad z_{abs} = 2.59$$

$$\frac{\Delta\mu}{\mu} = (2.06 \pm 0.79) \times 10^{-5}, \qquad z_{abs} = 3.02. \qquad (10.15)$$

Recently the $z = 3.02$ system has been reanalyzed [Wendt08], with the result that the claimed significance of Eq. (10.15) was not reproduced, and the absolute magnitude of the variation is bounded by $|\Delta\mu/\mu| \leq 4.9 \times 10^{-5}$ at 2σ, or

$$|\Delta\mu/\mu| \leq 2.5 \times 10^{-5}, \qquad z_{abs} = 3.02 \quad (1\sigma). \qquad (10.16)$$

Very recently, a new determination of the variation of μ appeared [King08] reporting a reanalysis of spectra from the same two H$_2$ absorption systems as [Reinhold06], and adding one additional system at $z \simeq 2.8$. The results of the new analysis are not consistent with the previous claim indicating a nonzero variation, either considering all three systems or the two previously considered. Instead, [King08] obtain a null bound, $\Delta\mu/\mu = (2.6 \pm 3.0) \times 10^{-6}$. As these results were published after this study has been finished, they are not considered any further.

The inversion spectrum of ammonia has been used to bound μ precisely at lower redshift [FlambaumNH3]. Recently the single known NH$_3$ absorber system at cosmological redshift has been analyzed [MurphyNH3], yielding

$$\frac{\Delta\mu}{\mu} = (0.74 \pm 0.89) \times 10^{-6}, \qquad z = 0.68. \qquad (10.17)$$

Tests of y

The 21cm HI line and molecular rotation spectra are sensitive to $y \equiv \alpha^2 g_p$, where g_p is the proton g-factor. Bounds on this quantity from [Murphy01] are

$$\frac{\Delta y}{y} = (-0.20 \pm 0.44) \times 10^{-5}, \qquad z = 0.247$$

$$\frac{\Delta y}{y} = (-0.16 \pm 0.54) \times 10^{-5}, \qquad z = 0.685. \qquad (10.18)$$

Tests of x

Further, the comparison of UV heavy element transitions with HI line probes for variations of $x \equiv \alpha^2 g_p \mu^{-1}$ [Tzanavaris06]: the weighted mean of nine analyzed systems yields

$$\frac{\Delta x}{x} = (0.63 \pm 0.99) \times 10^{-5}, \quad 0.23 < z_{abs} < 2.35. \qquad (10.19)$$

However, we note that i) the systems lie in two widely-separated low-redshift ($0.23 < z < 0.53$) and high-redshift ($1.7 < z < 2.35$) ranges; and ii) these two sub-samples have completely different scatter, χ^2/ν about the mean for the low- and high-redshift systems being 0.33, and 2.1, respectively. Hence we consider two samples, with average redshift $z = 0.40$ (5 systems) and $z = 2.03$ (4 systems). With expanded error bars in the high-redshift sample (after "method 3" of [Tzanavaris06]) we find

$$\frac{\Delta x}{x} = (1.02 \pm 1.68) \times 10^{-5}, \qquad \langle z \rangle = 0.40$$

$$\frac{\Delta x}{x} = (0.58 \pm 1.94) \times 10^{-5}, \qquad \langle z \rangle = 2.03. \qquad (10.20)$$

Tests of F

The comparison of HI and OH lines is sensitive to changes in $F \equiv g_p \left[\alpha^2 \mu\right]^{1.57}$ [Kanekar05] and yields

$$\frac{\Delta F}{F} = (0.44 \pm 0.36^{stat} \pm 1.0^{sys}) \times 10^{-5}, \quad z = 0.765. \tag{10.21}$$

Tests of F'

A similar method comparing CII and CO lines has very recently been proposed at high redshift [Levshakov07.2] yielding the best bound at redshifts > 4.5. The following bounds on $F' \equiv \alpha^2/\mu$ are obtained for two systems:

$$\frac{\Delta F'}{F'} = (0.1 \pm 1.0) \times 10^{-4}, \quad z = 6.42$$

$$\frac{\Delta F'}{F'} = (1.4 \pm 1.5) \times 10^{-4}, \quad z = 4.69. \tag{10.22}$$

10.4 The Oklo natural reactor

In Oklo/Gabon, a natural fission reactor formed by naturally enriched uranium in a rock formation with a water moderator was operating about 2 billion years ago ($\Delta t \simeq 1.8 \times 10^9$ y, $z \sim 0.14$ with WMAP5 best fit cosmology). The resulting isotopic ratios in this rock nowadays differ radically from any other terrestrial material. By modeling the nuclear fission process, one can in principle bound the variation of α over this period. The determination of $\Delta \ln \alpha$ at the time of the reactions results from considering the possible shift, due to variation of electromagnetic self-energy, in the position of a very low-lying neutron capture resonance of ^{149}Sm. The analysis of [Petrov05] gives the bound (taken as 1σ)

$$-5.6 \times 10^{-8} < \Delta\alpha/\alpha < 6.6 \times 10^{-8}. \tag{10.23}$$

For a linear time dependence this results in the bound

$$|\dot{\alpha}/\alpha| \leq 3 \times 10^{-17} \text{y}^{-1}. \tag{10.24}$$

Note that these results concern varying α only. If other parameters affecting nuclear forces, in particular light quark masses, are allowed to vary, the interpretation of this bound becomes unclear [Olive02, Flambaum02] since it depends on a nuclear resonance of ^{150}Sm whose properties are very difficult to investigate from first principles. In the absence of a resolution to this problem we consider Oklo as applying only to the α variation in each model. In scenarios where several couplings vary simultaneously we do not consider strong cancellations. Nevertheless, we allow for a certain degree of accidental cancellation and therefore multiply the error on the bound Eq. (10.23) by a factor three.

10.5 Meteorite dating

Meteorites which have formed at about the same time as the solar system, $t_{\text{Met}} \simeq 4.6 \times 10^9$ y ago ($z \simeq 0.44$) contain long-lived α- or β-decay isotopes. The decay rates of those isotopes may be sensitive probes of cosmological variation [Olive02, Olive03, Sisterna90]. Their (generally) small Q-values result from accidental cancellations between different contributions to nuclear binding energy, depending on fundamental couplings in different ways, thus the sensitivity of the decay rate may be enhanced by orders of magnitude.

The best bound concerns the ^{187}Re β-decay to osmium with $Q_\beta = 2.66$ keV. The decay rate λ_{187} is measured at present in the laboratory, and also deduced by isotopic analysis of meteorites formed about the same time as the solar system, 4.6×10^9 years ago. More precisely, the ratio λ_{187}/λ_U, averaged over the time between formation and the present, is measurable [Olive03, Fujii03], where λ_U is the rate of some other decay (for example uranium) used to calibrate meteorite ages.

The experimental values of λ_{187} imply (setting λ_U to a constant value)

$$t_{\text{Met}}^{-1} \int_{-t_{\text{Met}}}^{0} \frac{\Delta \lambda_{187}(t)}{\lambda_{187}} dt = 0.016 \pm 0.016. \tag{10.25}$$

Since the redshift back to t_{Met} is relatively small, we obtain bounds on recent time variation by assuming a linear evolution up to the present, for which the left hand side is $-(t_{\text{Met}}/2)\dot{\lambda}_{187}/\lambda_{187}$ and the fractional rate of change is bounded by

$$\frac{\dot{\lambda}_{187}}{\lambda_{187}} \simeq (-7.2 \pm 6.9) \times 10^{-12} \, \text{y}^{-1}. \tag{10.26}$$

Projected back to t_{Met} this gives the bound

$$\Delta \ln \lambda_{187} \simeq 0.033 \pm 0.032 \qquad (z \simeq 0.44). \tag{10.27}$$

This is a conservative bound unless the time variation has recently accelerated, or there are significant oscillatory variations over time.

Since the possible dependence of "control" decay rates λ_U/m_N on nuclear or fundamental parameters is much weaker than that of λ_{187}/m_N, we use this result for the variation of λ_{187} in units where λ_U is constant, i.e. $\Delta \ln(\lambda_{187}/\lambda_U) \simeq \Delta \ln(\lambda_{187}/m_N)$. We find the decay rate dependence to be [DSW08.1]

$$\Delta \ln \frac{\lambda_{187}}{m_N} \simeq -2.2 \times 10^4 \Delta \ln \alpha - 1.9 \times 10^4 \Delta \ln \frac{\hat{m}}{\Lambda_{QCD}} + 2300 \Delta \ln \frac{\delta_q}{\Lambda_{QCD}} - 580 \Delta \ln \frac{m_e}{\Lambda_{QCD}}. \tag{10.28}$$

10.6 Bounds on the variation of G_N

Variations of Newton's constant have been studied in the solar system and in astrophysical effects. Whilst all references give bounds exclusively on a potential variation of G_N, one should note that besides G_N also nuclear parameters (neutron / proton masses and parameters of nuclear forces) can vary, which would in general add degeneracies and make the results less stringent. It has generally been assumed that particle masses are constant, thus the resulting bounds actually constrain variation of $G_N m_N^2 \propto (m_N/M_P)^2$.

10.6. BOUNDS ON THE VARIATION OF G_N

In the solar system, changes of G_N induce changes in the orbits of planets. Range measurements to Mars from 1976 to 1982 can be used to obtain [Hellings83]

$$\dot{G}_N/G_N = 2 \pm 4 \times 10^{-12} \text{y}^{-1}. \tag{10.29}$$

Lunar laser ranging from 1970 to 2004 yields [Williams04]

$$\dot{G}_N/G_N = (4 \pm 9) \times 10^{-13} \text{y}^{-1}. \tag{10.30}$$

The stability of the orbital period of the binary pulsar PSR 1913+16 [Damour88] may be used to deduce

$$\dot{G}_N/G_N = (1.0 \pm 2.3) \times 10^{-11} \text{y}^{-1}. \tag{10.31}$$

All these results apply at the present epoch $z = 0$.

A bound on the behavior of G_N over the lifetime of the Sun (approximately 4.5×10^9y, $z = 0.43$) was found by Guenther et al. [Guenther98] by considering the effect of the resulting discrepancy in the helium/hydrogen fraction on p-mode oscillation spectra. The claimed constraint is

$$|\dot{G}_N/G_N| \leq 1.6 \times 10^{-12} \text{ y}^{-1}$$
$$|\Delta \ln G_N| \leq 7.2 \times 10^{-3} \quad z = 0.43, \tag{10.32}$$

where the assumed form of variation is a power law in time since the Big Bang, which may be approximated over the last few billion years as a linear dependence. For models with significantly nonlinear time dependence the bound may be reevaluated: since the bound arises from the accumulated effect of hydrogen burning since the birth of the Sun, it may be expressed as an integral of the variation over the Sun's lifetime analogous to Eq. (10.25).

The mass of neutron stars is determined by the Chandrasekhar mass

$$M_{\text{Ch}} \simeq \frac{1}{G_N^{3/2} m_n^2} \tag{10.33}$$

where m_n is the neutron mass. This may be reexpressed in terms of the baryon number of the star $n_B \propto M_{\text{Ch}}/m_n \propto (G_N m_n^2)^{-3/2}$, which is expected to be constant up to small corrections from matter accreting onto it. Thus the relative masses of neutron stars measured at the same epoch probes the fractional variation of $G_N m_n^2$ between their epochs of formation. From the comparison of masses of young and old neutron stars in binary systems (where the oldest neutron stars are up to 12 Gy old, $z \sim 3.3$), it is found [Thorsett96] that the variation of the average neutron star mass μ_n is $\dot{\mu}_n = -1.2 \pm 4.0(8.5) \times 10^{-12} M_\odot \text{ y}^{-1}$ at 60% (95%) confidence level. In units where particle masses are constant, we have

$$\dot{G}_N/G_N = -0.6 \pm 2.0\, (4.2) \times 10^{-12} \text{y}^{-1}, \tag{10.34}$$

where the averaging is performed over the last 12×10^9y, and the bound should be reinterpreted for variations which are not linear in time. The absolute variation over this period is then bounded at 1σ as

$$\Delta \ln G_N = (-0.7 \pm 2.4) \times 10^{-2}, \quad z = 3.3. \tag{10.35}$$

10.7 Atomic clocks

Atomic transitions can be measured in the laboratory to very high precision over periods of years. As each atomic transition depends differently on fundamental constants (*e.g.* α, μ), comparisons of different atomic transitions over long periods of time give very sensitive results.

Recently, stringent bounds on the present time variation of the fine structure constant and the electron-proton mass ratio have been obtained by [Blatt08],

$$d\ln\alpha/dt = (-0.31 \pm 0.3) \times 10^{-15}\,\text{y}^{-1}$$
$$d\ln\mu/dt = (1.5 \pm 1.7) \times 10^{-15}\,\text{y}^{-1}. \quad (10.36)$$

Fortier *et al.* [Fortier07] obtain stronger bounds, $|\dot\alpha/\alpha| < 1.3 \times 10^{-16}\,\text{y}^{-1}$, if other relevant parameters are assumed not to vary. If other atomic physics parameters are allowed to vary, this bound becomes considerably weaker, depending on a possible relative variation of the Cs magnetic moment and the Bohr magneton. Direct comparison of optical frequencies may yield bounds at the level of 10^{-17} per year; limits on variation of α from this method are reported with uncertainty $2.3 \times 10^{-17}\text{y}^{-1}$ [Rosenband08] but designated as preliminary. If these bounds are used then our limits from atomic clocks via α variation should be tightened by about an order of magnitude.

Extrapolating the results of [Blatt08] to the time of Oklo ($z = 0.14$, $t = 1.8 \times 10^9$ y) gives

$$\Delta\ln\alpha = (-0.56 \pm 0.54) \times 10^{-6},$$
$$\Delta\ln\mu = (-0.27 \pm 0.31) \times 10^{-5}. \quad (10.37)$$

Method	redshift	$\Delta \ln \alpha$ [10^{-6}]	$\Delta \ln \mu$ [10^{-5}]	$\Delta \ln G_N m_N^2$ [10^{-2}]	$\Delta \ln x$ [10^{-5}]	$\Delta \ln y$ [10^{-5}]	$\Delta \ln F$ [10^{-5}]	$\Delta \ln F'$ [10^{-4}]	$\Delta \ln \lambda_{187}$ [10^{-2}]
Oklo α [Petrov05]	0.14	0.00 ± 0.06							
21cm [Murphy01]	0.247					-0.20 ± 0.44			
Sun [Guenther98]	0.43			0 ± 0.72					
Heavy/HI, low-z [Tzanavaris06]	0.40				1.0 ± 1.7				
Meteorite [Olive03]	0.44								3.3 ± 3.2
Mα epoch 2 [Murphy03.2]	0.65	-2.9 ± 3.1							
Ammonia [FlambaumNH3]	0.68		0.06 ± 0.19						
21cm [Murphy01]	0.685					-0.16 ± 0.54			
HI / OH [Kanekar05]	0.765						0.4 ± 1.1		
Absorption [Fujii07]	1.15	-0.1 ± 1.8							
Mα epoch 3 [Murphy03.2]	1.47	-5.8 ± 1.3							
Absorption [Levshakov07.1]	1.84	5.7 ± 2.7							
Heavy/HI, high-z [Tzanavaris06]	2.03				0.6 ± 1.9				
H_2 [Reinhold06]	2.59		2.78 ± 0.88						
Mα epoch 4 [Murphy03.2]	2.84	-8.7 ± 3.7							
H_2 [Reinhold06]	3.02		2.06 ± 0.79						
Neutron stars [Thorsett96]	3.3				-0.7 ± 2.4				
CII / CO [Levshakov07.2]	4.69							1.4 ± 1.5	
CII / CO [Levshakov07.2]	6.42							0.1 ± 1.0	
CMB [Martins03], [Chan07]	10^3	$0^{+1 \times 10^4}_{-3 \times 10^4}$			0^{+7}_{-6}				

Table 10.1: Observational 1σ bounds on variations. Observables are defined as $\mu \equiv m_p/m_e$, $x \equiv \alpha^2 g_p \mu^{-1}$, $y \equiv \alpha^2 g_p$, $F \equiv g_p[\alpha^2 \mu]^{1.57}$, $F' \equiv \alpha^2/\mu$. The given redshift may denote a single measurement, or an averaged value over a certain range: see main text. The two CMB bounds are independent of each other. Our BBN bounds cannot be displayed in this form.

Chapter 11

Variations from BBN to today in unified scenarios

As has been laid out in Sec. 4.5, we will use the concept of grand unification to reduce the number of potentially varying parameters. The main idea is that the GUT relations interrelate variations of the "fundamental" parameters G_k which we defined in Sec. 8.1.

In this chapter, we consider the hypothesis that, for all redshifts, all fractional variations in the "fundamental" parameters G_k are proportional to one nontrivial variation with fixed constants of proportionality. If the variation of the unified gauge coupling $\Delta \ln \alpha_X$ is nonvanishing, we may write

$$\Delta \ln G_k = d_k \Delta \ln \alpha_X \qquad (11.1)$$

for some constants d_k, assuming small variations. Different unification scenarios correspond to different sets of values for the "unification coefficients" d_k. Considering the values of $\Delta \ln G_k$ as coordinates in an N_k-dimensional space, this assumption restricts variations to a single line passing through zero. The variation then constitutes exactly one degree of freedom. We will go beyond this hypothesis in the next chapter where we also consider models with growing neutrinos and oscillating variations (see Sec. 5.4) for which a fixed linear relation (11.1) is not realized for all z.

11.1 GUT relations

GUT relations have the property that variations of the Standard Model gauge couplings and mass ratios can be determined in terms of a smaller set of parameters describing the unified theory and its symmetry breaking. Hence, if nonzero variations in different observables are measured at similar redshifts, models of unification may be tested without referring to any specific hypothesis for the overall cosmological history of the variation. We need only assume that for a given range of z the time variation is slow and approximately homogeneous in space, hence $\Delta \ln \alpha_X$ depends only on redshift z to a good approximation. The relevant unified parameters are the unification mass M_{GUT} (relative to the Planck mass), the GUT coupling α_X defined at the scale M_{GUT}, the Higgs v.e.v. $\langle \phi \rangle$ and, for supersymmetric theories, the soft supersymmetry breaking masses $\tilde{m}$, which enter in the renormalization group (RG)

11.1. GUT RELATIONS

equations for the running couplings. Then, for the variations at any given z we can write

$$\Delta \ln \frac{M_{GUT}}{M_P} = d_M l, \quad \Delta \ln \alpha_X = d_X l, \quad \Delta \ln \frac{\langle\phi\rangle}{M_{GUT}} = d_H l, \quad \Delta \ln \frac{\tilde{m}}{M_{GUT}} = d_S l, \quad (11.2)$$

where $l(z)$ is the "evolution factor" introduced for later convenience. If α_X varies nontrivially we may normalize l via $d_X = 1$. In supersymmetric theories we set $\alpha_X = 1/24$, in nonsupersymmetric theories we set $d_S \equiv 0$ and $\alpha_X = 1/40$ (see Sec. 4.5).

We make the simplifying assumption that the masses of Standard Model fermions all vary as the Higgs v.e.v., *i.e.* Yukawa couplings are constant at the unification scale:

$$\Delta \ln \frac{m_e}{M_{GUT}} = \Delta \ln \frac{\delta_q}{M_{GUT}} = \Delta \ln \frac{\hat{m}}{M_{GUT}} = \Delta \ln \frac{m_s}{M_{GUT}} = \Delta \ln \frac{\langle\phi\rangle}{M_{GUT}}. \quad (11.3)$$

Like the gauge couplings, also the fermion masses vary under variations of the unified coupling α_X due to the renormalization group running of fermion masses. However, we have explicitly calculated the effect of varying couplings and found that it is at the order of a 1% correction[1], which is already smaller than our uncertainties in hadronic and nuclear physics[2]. Hence we can apply the assumption (11.3). Using the relations (4.23) and (4.26), one finds for the QCD scale

$$\frac{\Delta \ln(\Lambda_{QCD}/M_{GUT})}{l} = \frac{2\pi}{9\alpha_X} d_X + \frac{2}{9} d_H + \frac{4}{9} d_S \quad (11.4)$$

and for the fine structure constant,

$$\frac{\Delta \ln \alpha}{l} = \frac{80\alpha}{27\alpha_X} d_X + \frac{43}{27} \frac{\alpha}{2\pi} d_H + \frac{257}{27} \frac{\alpha}{2\pi} d_S. \quad (11.5)$$

For the nucleon mass we include possible strange quark contributions. In our treatment of BBN, we neglected strange quark contributions, as the final dependence on m_s was much below the model uncertainty. Here we include the roughly known strange contribution to the nucleon mass. The uncertainty in the strangeness content is an indicator of the overall uncertainty that may arise due to m_s variation. We found (Eqs. (8.20), (8.21) and (8.14))

$$\Delta \ln \frac{m_N}{\Lambda_{QCD}} = 0.048 \Delta \ln \frac{\hat{m}}{\Lambda_{QCD}} + (0.12 \pm 0.12) \Delta \ln \frac{m_s}{\Lambda_{QCD}}, \quad (11.6)$$

$$\Delta \ln \frac{Q_N}{\Lambda_{QCD}} = -0.59 \Delta \ln \alpha + 1.59 \Delta \ln \frac{\delta_q}{\Lambda_{QCD}}, \quad (11.7)$$

and thus

$$\frac{\Delta \ln \mu}{l} = (0.58 \mp 0.08) \frac{d_X}{\alpha_X} + (0.37 \mp 0.05) d_S + (-0.65 \pm 0.09) d_H, \quad (11.8)$$

$$\frac{\Delta \ln(G_N m_N^2)}{l} = 2d_M + (1.16 \mp 0.17) \frac{d_X}{\alpha_X} + (0.74 \mp 0.11) d_S + (0.71 \pm 0.19) d_H, \quad (11.9)$$

---

[1] For low-energy observables such as $m_q(Q^2)/\Lambda_{QCD}$ we consider an RG scale Q^2 that is fixed relative to Λ_{QCD}, $Q^2 = \text{const}\cdot\Lambda_{QCD}^2$. Thus the variation of $m_q(Q^2)/m_q(M_{GUT}^2)$ is entirely due to the dependence on $\alpha_3(M_{GUT})$, which is suppressed by a loop factor α_X/π compared to the nonperturbative dependence of Λ_{QCD}/M_{GUT} on α_X. We find $\Delta \ln(\bar{m}_q(Q^2)/\bar{m}_q(M_{GUT}^2)) = 2/7 \Delta \ln \alpha_X \simeq (9\alpha_X/7\pi)\Delta \ln(\Lambda_{QCD}/M_{GUT})$ under variation of α_X, where $\bar{m}_q$ is the running quark mass.
[2] Langacker *et al.* [Langacker01] arrive at the same conclusion.

where the upper or lower signs correspond to the positive or negative signs in Eq. (11.6) respectively.

The largest contribution to variations of the proton g-factor g_p has been argued to arise from the pion loop [Murphy03.2], yielding at first order a dependence on the light quark mass of

$$\Delta \ln g_p \simeq -0.087 \Delta \ln \hat{m}/\Lambda_{QCD},$$
$$\frac{\Delta \ln g_p}{l} \simeq 0.06 \frac{d_X}{\alpha_X} - 0.07 d_H + 0.04 d_S. \tag{11.10}$$

Hence the variations of observables including g_p are

$$\frac{\Delta \ln x}{l} = (-0.48 \pm 0.08)\frac{d_X}{\alpha_X} + (0.59 \mp 0.09)d_H + (-0.31 \pm 0.05)d_S$$
$$\frac{\Delta \ln y}{l} = 0.10 \frac{d_X}{\alpha_X} - 0.06 d_H + 0.06 d_S$$
$$\frac{\Delta \ln F}{l} = (1.04 \mp 0.13)\frac{d_X}{\alpha_X} + (-1.08 \pm 0.14)d_H + (0.65 \mp 0.08)d_S$$
$$\frac{\Delta \ln F'}{l} = (-0.54 \pm 0.08)\frac{d_X}{\alpha_X} + (0.65 \mp 0.09)d_H + (-0.35 \pm 0.05)d_S. \tag{11.11}$$

We have now expressed the variations accessible to observation in terms of three (four) variables: l, d_X, d_H (and d_S), where one parameter may be eliminated by normalization. Different unified scenarios will be characterized by different relations among these parameters.

11.2 Variations in six different unified scenarios

We will now investigate six different scenarios for the variation of the grand unified parameters α_X, M_{GUT}/M_P, $\langle \phi \rangle / M_{GUT}$ and $\tilde{m}/M_{GUT}$. These will fix the unification coefficients d_k. For each unified scenario we display the z-dependence of the fractional variation. Each figure shows the available information from observations of different couplings, interpreted as constraints on the variation of a single parameter. Since we have only one free variable we can plot all observations simultaneously as a function of redshift. Inspection "by eye" permits to judge if a smooth and monotonic evolution of the varying parameter is consistent or not. Most data points are upper bounds on a possible variation, and for the non-zero variations we can study two immediate questions in each scenario:

First, whether claimed nonzero variations of α [Murphy03.2] and μ [Reinhold06] at redshift 2–3 are compatible with one another, since the ratio of their fractional variations is predicted in each scenario.

Second, we consider whether there is an indication of nonzero variation at BBN. For no variation at BBN we obtain $\chi^2 = 17.9$ for 3 measured abundances (^{4}He, D, ^{7}Li). This discrepancy between theory and observation is exclusively due to ^{7}Li. (Considering only ^{4}He and D, the value of χ^2 is 0.24.) If we wish to solve or ameliorate the "lithium problem" by a nonzero variation, we will require χ^2/ν to be not much larger than unity, taking $\nu = 2$ as appropriate for one adjustable parameter. If there is no significant range where the three abundances have a 2σ fit ($\chi^2/\nu \leq 4$) then we give up the hypothesis that the ^{7}Li problem is solved by coupling variations and instead assume that the observed depletion is due to some astrophysical effect. In

11.2. VARIATIONS IN SIX DIFFERENT UNIFIED SCENARIOS

this case we consider only D and ^{4}He abundances as observational bounds on the size of variations at BBN.

11.2.1 Varying α alone

Before describing the six different grand unified scenarios, we consider a variation of the fine structure constant α alone. Clearly here we are unable to account for any nonzero variation in μ or other quantities independent of α. The cosmological history is dominated by the nonzero variation of the Mα values at redshifts $z \simeq 1$ to 4. We find that there is almost no 2σ match of the BBN values ($\chi^2/\nu \geq 3.9$): the 2-sigma range is

$$3.25\% \geq \Delta \ln \alpha_{BBN} \geq 4.06\%. \tag{11.12}$$

Hence it seems unlikely that the "lithium problem" can be solved by a variation of α alone. If we regard the ^{7}Li discrepancy as due to systematic or astrophysical effects we can set a conservative bound on α variation from ^{4}He and D abundances

$$-3.6\% \geq \Delta \ln \alpha_{BBN} \geq 1.9\%, \tag{11.13}$$

where we imposed that neither the D nor ^{4}He abundance should deviate by more than 2σ from observational values. See Fig. 11.1 for a summary of the bounds in this case.

11.2.2 Scenario 1: Varying gravitational coupling

In this scenario we have only d_M nonvanishing,

$$d_H = d_S = d_X = 0, \tag{11.14}$$

therefore

$$\Delta \ln \frac{M_{GUT}}{M_P} = \frac{1}{2} \Delta \ln G_N \Lambda_{QCD}^2. \tag{11.15}$$

We find that there is no value of $\Delta \ln G_N \Lambda_{QCD}^2$ for which BBN is consistent with the three observed abundances within 2σ. The best fit values are $\chi^2/\nu \geq 7.7$ for no variation of m_N/M_P at CMB and $\chi^2/\nu \geq 5.9$ if the variation of m_N/M_P has the same size at BBN and CMB. Assuming that the discrepancy in the ^{7}Li abundance is due to some other effect, we find the allowed region of variation of G_N at BBN under which primordial D and ^{4}He abundance lie within the observed range at 1σ (2σ),

$$-5\% \, (-13\%) \leq \Delta \ln G_N \Lambda_{QCD}^2 \leq 12\% \, (22\%) \tag{11.16}$$

If the variation of m_N/M_P has the same size at BBN and CMB one finds

$$-4\% \, (-11\%) \leq \Delta \ln G_N \Lambda_{QCD}^2 \leq 10\% \, (16\%). \tag{11.17}$$

The bounds on time variation of $G_N \Lambda_{QCD}^2$ are much weaker than for many other varying couplings. This scenario also predicts a vanishing value of η in Eötvös experiments (see Sec. 12.3 for details). Thus, to any one of the following scenarios we may add an additional nonzero d_M of similar size to d_X, d_H or d_S without changing the results significantly.

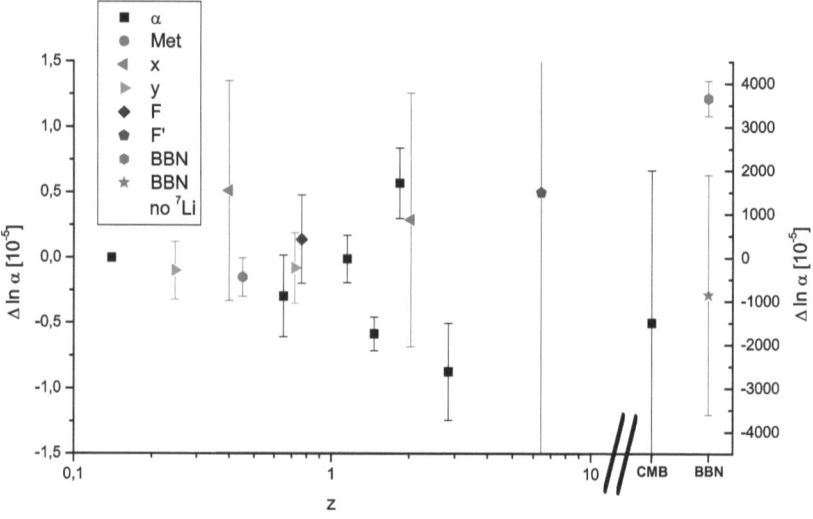

Figure 11.1: Variations for varying α alone. Only observations constraining α variation are shown; the BBN fit including ^{7}Li is poor ($\chi^2/\nu \geq 7.8/2$) hence we also display a conservative bound from ^{4}He and D abundances neglecting ^{7}Li.

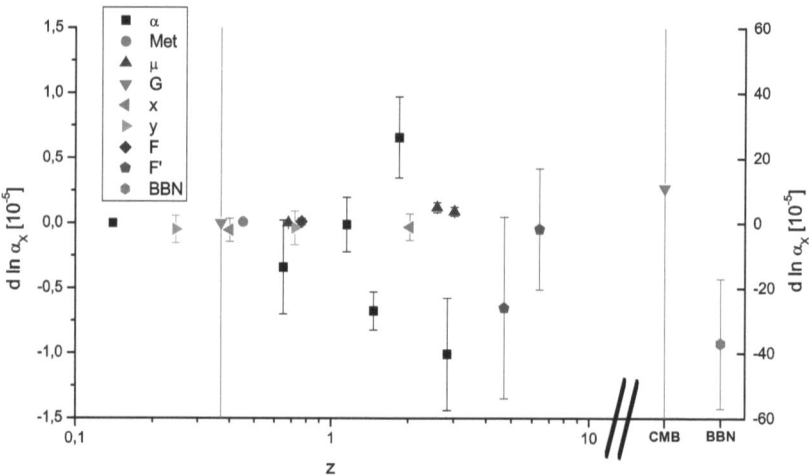

Figure 11.2: Variations for scenario 2; BBN bounds are 2σ bounds.

11.2. VARIATIONS IN SIX DIFFERENT UNIFIED SCENARIOS

11.2.3 Scenario 2: Varying unified coupling

In the first GUT scenario without SUSY we consider the case when only d_X is non-vanishing,
$$d_H = d_S = d_M = 0, \qquad \alpha_X = 1/40. \tag{11.18}$$
Within a supersymmetric theory the same relations will apply except that $\alpha_X = 1/24$ and the variations of observables are scaled by a factor $24/40$ relative to $\Delta \ln \alpha_X$: we designate this as Scenario 2S. In both cases we find here
$$\frac{\Delta \ln \mu}{\Delta \ln \alpha} = 27. \tag{11.19}$$
It is then highly unlikely for the nonzero $M\alpha$ result for variation of α to coexist with the determination of μ at redshift around 3 [Reinhold06], even if the latter is interpreted as an upper bound on the absolute size of variation [Wendt08].

For the BBN fit, we find without SUSY (excluding modifications of the baryon fraction η due to varying m_N) no range of values fitting at 1σ level ($\chi^2/\nu \geq 2.3$). At 2σ the abundances, including ^{7}Li, become consistent for the range
$$-5.7 \times 10^{-4} \leq \Delta \ln \alpha_X \leq -1.7 \times 10^{-4} \qquad (2\sigma). \tag{11.20}$$
If one includes a variation of m_N at the time of CMB with the same magnitude as at BBN the result remains unchanged ($\chi^2/\nu \geq 2.45$), with the same 2σ range. For this scenario we may consider a nonzero variation at BBN, but more recent probes must all be viewed as increasingly tight null bounds.

11.2.4 Scenario 3: Varying Fermi scale

In this scenario we consider the case when the variation arises solely from a change in the Higgs expectation value relative to the unified scale, thus only d_H is nonzero:
$$d_S = d_M = d_X = 0, \qquad \alpha_X = 1/40. \tag{11.21}$$
This scenario implies
$$\frac{\Delta \ln \mu}{\Delta \ln \alpha} = -325. \tag{11.22}$$
Whether we interpret the determination of μ [Reinhold06] as a detection or an upper bound, any variation in α at large redshift should be orders of magnitude smaller than current observational sensitivity.

We find for BBN including ^{7}Li ($\nu = 2$) no 1σ range ($\chi^2/\nu \geq 1.95$) but
$$6 \times 10^{-3} \leq \Delta \ln \langle \phi \rangle / M_{GUT} \leq 22 \times 10^{-3} \qquad (2\sigma). \tag{11.23}$$
A variation of m_N at the time of CMB with the same magnitude as at BBN does not change this result.

11.2.5 Scenario 4: Varying Fermi scale and SUSY-breaking scale

This scenario corresponds to scenario 3, but includes supersymmetry and assumes that the mass-generating mechanism for SM particles and their superpartners gives rise to the same variation:
$$d_M = d_X = 0, \qquad d_S = d_H, \qquad \alpha_X = 1/24. \tag{11.24}$$

We find here
$$\frac{\Delta \ln \mu}{\Delta \ln \alpha} = -21.5, \tag{11.25}$$
such that again the claimed nonzero variations in α and μ cannot be compatible and the variation in α at redshift 3 must be below current sensitivities. We demonstrate this in Fig. 11.4, where we show for this scenario the bounds on the variable $d_H l = \Delta \ln(\langle\phi\rangle/M_{GUT})$ that arise from various observations.

We find for BBN including ^{7}Li ($\nu = 2$) no 1σ fit ($\chi^2/\nu \geq 1.60$), while at 2σ
$$1.25 \times 10^{-2} \leq \Delta \ln\langle\phi\rangle/M_{GUT} \leq 5.4 \times 10^{-2} \quad (2\sigma). \tag{11.26}$$

If one includes a variation of m_N at the time of CMB with the same magnitude as at BBN the allowed range becomes slightly restricted ($\chi^2/\nu \geq 1.72$),
$$1.20 \times 10^{-2} \leq \Delta \ln\langle\phi\rangle/M_{GUT} \leq 4.9 \times 10^{-2} \quad (2\sigma). \tag{11.27}$$

11.2.6 Scenario 5: Varying unified coupling and Fermi scale

In this scenario we study a combined variation of the unified coupling and the Higgs expectation value:
$$d_M = d_S = 0, \quad d_H = \tilde{\gamma} d_X, \quad \alpha_X = 1/40. \tag{11.28}$$

The parameter $\tilde{\gamma}$ can be related to the parameter $\gamma \equiv \frac{\Delta \ln\langle\phi\rangle/M_{GUT}}{\Delta \ln \Lambda_{QCD}/M_{GUT}}$ which was introduced in Sec. 9.2 via
$$\gamma = \tilde{\gamma}\left(\frac{2\pi}{9\alpha_X} + \frac{2}{9}\tilde{\gamma}\right)^{-1}. \tag{11.29}$$

There we examined the cases $\gamma = (0, 1, 1.5)$ which correspond to $\tilde{\gamma} = (0, 36, 63)$. Here we find that the best BBN fit is reached for $\tilde{\gamma} \approx 50$ with $\chi^2/\nu = 1.45$. Note that we have the freedom to adjust $\tilde{\gamma}$ such that nonzero variations of α and μ at redshift $\simeq 3$ are consistent with each other. We have
$$\frac{\Delta \ln \mu}{\Delta \ln \alpha} = \frac{23.2 - 0.65\tilde{\gamma}}{0.865 + 0.002\tilde{\gamma}}. \tag{11.30}$$

We choose for illustration $\tilde{\gamma} = 42$, for which
$$\Delta \ln \mu = -5.6 \, \Delta \ln \alpha \tag{11.31}$$
and the 2σ contour for BBN is
$$7.5 \times 10^{-4} \leq \Delta \ln \alpha_X \leq 28 \times 10^{-4}. \tag{11.32}$$

For a variation of m_N at the time of CMB with the same magnitude as at BBN the fit becomes worse ($\chi^2/\nu \geq 1.68$). However, a 2σ fit to BBN is obtained over a wide range of $0 \leq \tilde{\gamma} \leq 26$ (negative $\Delta \ln \alpha_X$) and $40 \leq \tilde{\gamma} < \infty$ (positive $\Delta \ln \alpha_X$).

Assuming that the apparent ^{7}Li mismatch at BBN is due to systematic astrophysical effects, we may bound α_X with only D and ^{4}He abundances. Here we find at 1σ
$$-5.5 \times 10^{-4} \leq \Delta \ln \alpha_X \leq 1.44 \times 10^{-3}. \tag{11.33}$$

In Fig. 11.5 we again plot simultaneously all observations for this scenario. This shows that the bound from BBN including ^{7}Li is not consistent with the claimed nonzero variations of α and μ for a monotonic evolution over z.

11.2. VARIATIONS IN SIX DIFFERENT UNIFIED SCENARIOS

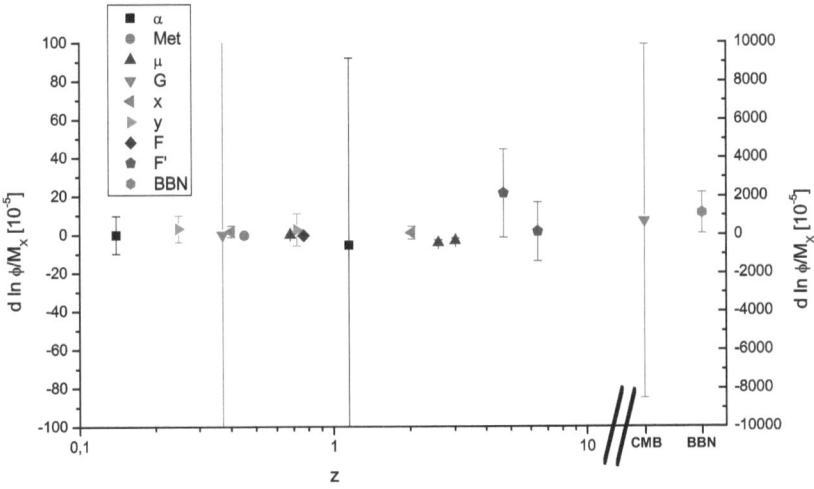

Figure 11.3: Variations for scenario 3; BBN bounds are 2σ. Note that due to the very large ratio $\Delta\ln\mu/\Delta\ln\alpha$ in this scenario, points indicating any nonzero variation of α fall well outside the range of the graph.

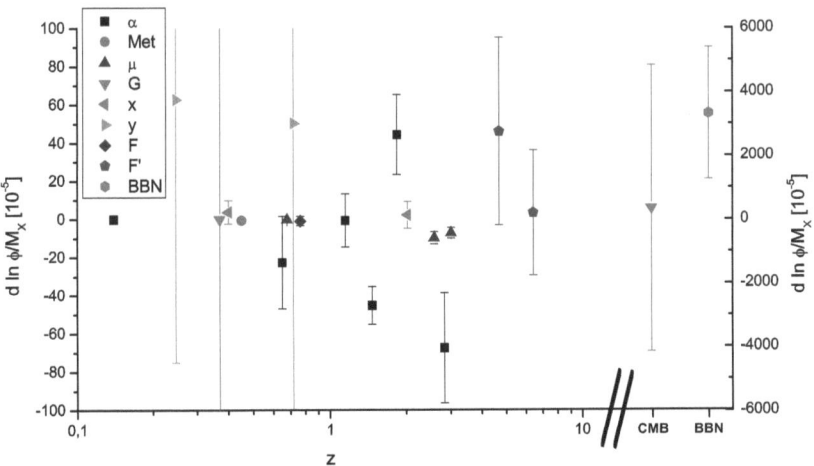

Figure 11.4: Variations for scenario 4; BBN bounds are 2σ

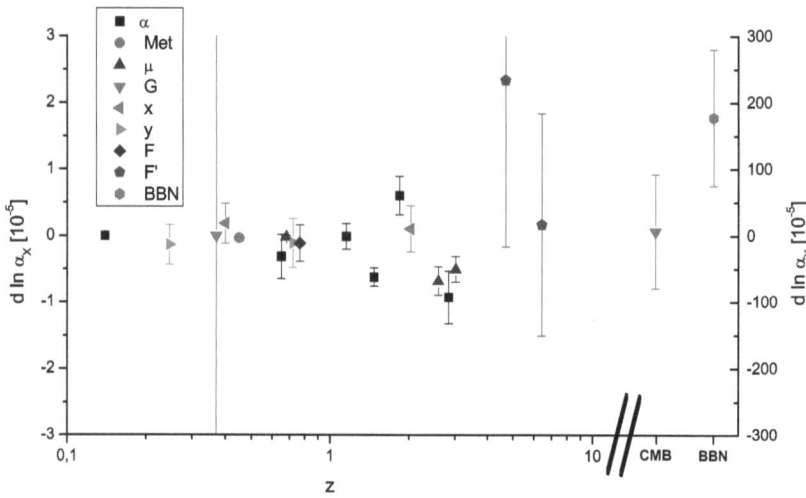

Figure 11.5: Variations for scenario 5, $\tilde{\gamma} = 42$; BBN bounds are 2σ

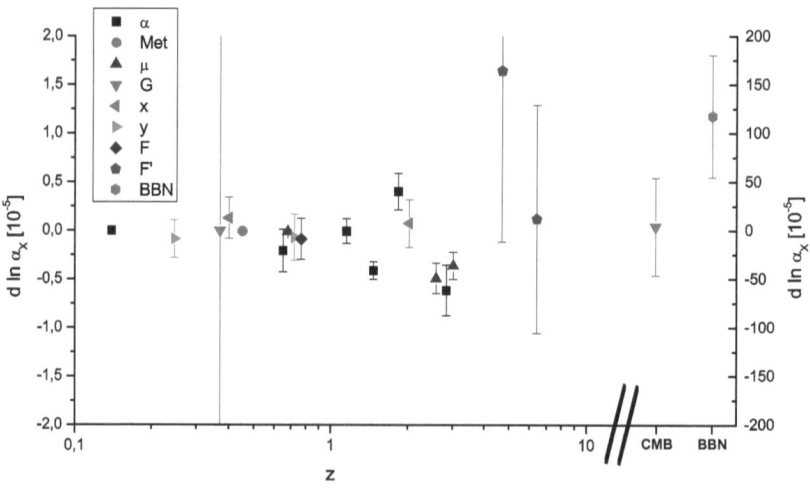

Figure 11.6: Variations for scenario 6, $\tilde{\gamma} = 70$; BBN bounds are 2σ

11.2. VARIATIONS IN SIX DIFFERENT UNIFIED SCENARIOS

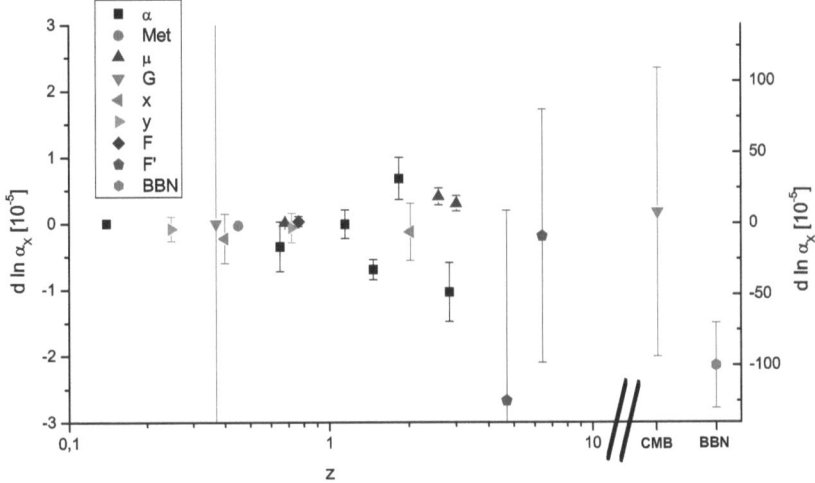

Figure 11.7: Variations for scenario 6, $\tilde{\gamma} = 25$; BBN bounds are 2σ

11.2.7 Scenario 6: Varying unified coupling and Fermi scale with SUSY

In this scenario we study a combined variation of the unified coupling and the Higgs v.e.v. including SUSY, where as in Scenario 4 we tie the variations of the superpartner masses and Fermi scale together:

$$d_M = 0, \qquad d_S \simeq d_H = \tilde{\gamma} d_X, \qquad \alpha_X = 1/24. \tag{11.34}$$

Now the relation to γ is modified as

$$\gamma = \tilde{\gamma}\left(\frac{2\pi}{9\alpha_X} + \frac{2}{3}\tilde{\gamma}\right)^{-1}. \tag{11.35}$$

One may again adjust $\tilde{\gamma}$ to make nonzero variations in α and μ self-consistent, where now

$$\frac{\Delta \ln \mu}{\Delta \ln \alpha} = \frac{14 - 0.28\tilde{\gamma}}{0.52 + 0.013\tilde{\gamma}}. \tag{11.36}$$

We find that a good fit to BBN is obtained over a large range of $\tilde{\gamma}$, ranging from $\tilde{\gamma} = 100$ to infinity with minimal $\chi^2/\nu = 1.45$. This shows that the main effect in the SUSY model comes from the variation of the Higgs v.e.v. Including a variation of m_N at the time of CMB with the same magnitude as at BBN the fits gets worse ($\chi^2/\nu \geq 1.8$). A 2σ fit can be obtained for $0 \leq \tilde{\gamma} \leq 28$ (for negative $\Delta \ln \alpha_X$ at BBN) and for $58 \leq \tilde{\gamma} < \infty$ (positive $\Delta \ln \alpha_X$).

First, we study the case $\tilde{\gamma} = 70$ for which

$$\Delta \ln \mu = -3.9 \Delta \ln \alpha \qquad (\tilde{\gamma} = 70) \tag{11.37}$$

and BBN is fit with a 2σ range
$$5.5 \times 10^{-4} \leq \Delta \ln \alpha_X \leq 18 \times 10^{-4}. \tag{11.38}$$
Neglecting ^{7}Li, we obtain a 1σ bound from BBN
$$-3.5 \times 10^{-4} \leq \Delta \ln \alpha_X \leq 9.3 \times 10^{-4}. \tag{11.39}$$
Secondly, we study the case $\tilde{\gamma} = 25$ where
$$\Delta \ln \mu = 8.3 \Delta \ln \alpha \qquad (\tilde{\gamma} = 25), \tag{11.40}$$
and where the 2σ contour for BBN is
$$-13 \times 10^{-4} \leq \Delta \ln \alpha_X \leq -7 \times 10^{-4}. \tag{11.41}$$
In this second case the Murphy α measurement and BBN point into the same direction. The difference between the two values of $\tilde{\gamma}$ can be seen from a comparison of Figs. 11.6 and 11.7.

11.3 Epochs and evolution factors

11.3.1 Epochs

As a next step, we group the information on experimental bounds on variations of couplings in the different unified scenarios (displayed in Figs. 11.1 to 11.7) into different cosmological epochs. This produces a first quantitative estimate of the possible time evolution for the various unified scenarios. The choice of epochs is somewhat arbitrary. Two epochs are singled out by events in early cosmology, namely the last scattering surface of CMB, and BBN. The very recent epoch comprises present day laboratory experiments and the Oklo natural reactor, for which a linear interpolation to the present rate of varying couplings seems reasonable. We further divide the observations at intermediate redshift into three epochs.

- **Epoch 1:** Today until Oklo
  Contains Oklo and laboratory measurements. For the laboratory measurements, we extrapolate the rate of change of the couplings to finite changes at the redshift $z = 0.14$ ($t = 1.8 \times 10^9$ y) of the Oklo event.

- **Epoch 2:** $0.2 \leq z \leq 0.8$
  Contains absorption spectra and isotopic abundance measurements in meteorites. We chose a boundary $z = 0.8$ since the Murphy dataset [Murphy03.2] has relatively few systems around this redshift, making a natural division.

- **Epoch 3:** $0.8 \leq z \leq 2.4$
  Contains several absorption spectra measurements. The end of the Tzanavaris dataset [Tzanavaris06] sets the cut at $z = 2.4$.

- **Epoch 4:** $2.4 \leq z \leq 10$
  Contains absorption spectra measurements and bounds on G_N from neutron stars.

- **Epoch 5:** CMB, $z \approx 1100$

- **Epoch 6:** BBN, $z \approx 10^{10}$

11.3. EPOCHS AND EVOLUTION FACTORS

11.3.2 Evolution factors

We define "evolution factors" l_n for epochs $n = 1, \ldots, 6$ by

$$\Delta \ln G_{k,n} = d_k l_n. \tag{11.42}$$

For each unification scenario we proceed to a quantitative estimate of l_n, shown in Tab. 11.1. The usefulness of considering the evolution factors l_n is that the unknown (and possibly not monotonic) behavior of the mechanism driving the coupling variations is rolled into a finite number of parameters. For a monotonic behavior they satisfy $l_n < l_p$ whenever $z_n < z_p$. The basic assumption remains the proportionality $\Delta \ln G_k(z_n) = d_k l(z_n) = d_k l_n$, with constant unification coefficients d_k independent of the epoch. The normalization of l_n is arbitrary, and we take for scenarios 2, 5 and 6

$$l_n = \Delta \ln \alpha_{X,n}, \tag{11.43}$$

while for scenarios 3 and 4 we take

$$l_n = \Delta \ln(\langle \phi \rangle / M_{GUT}),_n. \tag{11.44}$$

For each epoch and scenario, we compute the evolution coefficients l_n as a weighted average over the measurements in the epoch. The representative redshift z_n is the average over the redshifts of observations inside the corresponding epoch. It is shown together with the resulting values for l_n in Tab. 11.1. This table summarizes our results under the assumption of proportionality.

Rates of time variation in the present epoch

For epoch 1 we incorporate the laboratory measurements for rates of varying couplings by linear extrapolation in time to the Oklo redshift $z_1 = 0.14$. The logarithmic time derivatives may be approximated by linear interpolation

$$\frac{\dot{G}_k}{G_k} = \partial_t \ln G_k \simeq -\frac{d_k l_1}{t_0 - t_1}, \tag{11.45}$$

where $t_1 = 1.8 \times 10^9 \text{y}$ is the time corresponding to the redshift $z_1 = 0.14$.

Method of averaging

We evaluate the weighted average using all values listed in Tab. 10.1. This procedure may be quite problematic, since sometimes different observations are in manifest contradiction. We take the attitude that, given the possible presence of systematic effects both in spectroscopic determinations of nonzero coupling variations and in the primordial ^{7}Li abundance, a viable model need not fit all data points. However, even if any given nonzero claimed variation is actually due to systematic error, we still expect the size of the error to be comparable to the size of the claimed variation. Thus, such claims are most conservatively interpreted as bounds on the absolute magnitude of variation. The surviving nonzero variation(s), in addition to the null bounds at other epochs, define a set of evolution factors which must be satisfied by any explicit model of evolution.

For some scenarios we therefore also evaluate the evolution factors that are obtained by considering that some of the claimed observations of nonzero variation may instead be due to an underestimated systematic error. These alternative evolution factors are given in square brackets, corresponding to the following replacements:

Epoch	1	2	3	4	5	6
z_n	0.14	0.53	1.6	3.8	10^3	10^{10}
Scenario	$l_1 \times 10^6$	$l_2 \times 10^6$	$l_3 \times 10^5$	$l_4 \times 10^5$	$l_5 \times 10^4$	$l_6 \times 10^3$
α only	-0.01 ± 0.06	-1.1 ± 1.0	-0.26 ± 0.10	-0.85 ± 0.37	-150 ± 350	5 ± 34
2	-0.1 ± 0.1	0.04 ± 0.03	-0.15 ± 0.08	0.10 ± 0.03	0.9 ± 14	-0.37 ± 0.20
3	4.1 ± 4.8	-1.5 ± 1.2	0.42 ± 3.3	-3.6 ± 0.9	69 ± 920	14 ± 8
4	3.9 ± 8.5	-3.4 ± 2.7	-8.4 ± 5.1	-8.7 ± 2.1	31 ± 450	33 ± 21
5, ($\tilde{\gamma} = 42$)	-0.02 ± 0.18	-0.24 ± 0.18	-0.25 ± 0.10	-0.61 ± 0.13	0.6 ± 8.6	1.7 ± 1.1 [0.4 ± 1.0]
6, ($\tilde{\gamma} = 70$)	-0.02 ± 0.12	-0.10 ± 0.07	-0.17 ± 0.07	-0.44 ± 0.10	0.3 ± 5.0	1.2 ± 0.6 [0.3 ± 0.6]
6, ($\tilde{\gamma} = 25$)	-0.12 ± 0.18	0.04 ± 0.12	-0.30 ± 0.11	0.29 ± 0.08 [-0.43 ± 0.28]	0.7 ± 10	-1 ± 0.3

Table 11.1: Redshifts and evolution factors for each epoch, for the scenarios defined in Sec. 11.2. In the first row the values of l_n give the fractional variation of α; in Scenarios 2, 5 and 6 that of α_X; and in 3 and 4 that of $\langle\phi\rangle/M_{GUT}$. Values in brackets give, for BBN (l_6) the evolution factors neglecting ^{7}Li; or for l_4, the evolution factor with the $\Delta\mu/\mu$ value of [Reinhold06] substituted by that of [Wendt08].

Scenario 5, $\tilde{\gamma} = 42$: Neglecting ^{7}Li-abundance at BBN
Scenario 6, $\tilde{\gamma} = 70$: Neglecting ^{7}Li-abundance at BBN
Scenario 6, $\tilde{\gamma} = 25$: Replacing the μ measurements of [Reinhold06] by the conservative upper bound of [Wendt08].
In the case where α alone varies, since the fit including ^{7}Li is poor we calculate a 2σ range using observational central values and errors of D and ^{4}He abundances as explained in Sec. 11.2.1.

11.3.3 Monotonic evolution with unification

It seems natural to expect that variations of constants, if they occurred, evolve monotonically[3]. Looking on Figs. 11.1 to 11.7 and Tab. 11.1, we can ask the question whether the claimed variations are consistent with monotonic variation within the specific GUT scenarios. Here we briefly summarize whether the unified scenarios we consider can be consistent with a monotonic evolution of the single underlying varying parameter.

Varying α only

Although variation of α alone does not help to account for deviation of BBN abundances from standard theory, or for any nonzero variation of μ, the cosmic history is interesting due to the significant nonzero value in Epochs 3 and 4. The Oklo bound in Epoch 1 restricts the present time variation to $3.7 \times 10^{-17}\,\mathrm{y}^{-1}$ (assuming no acceleration of $\partial_t \alpha$).

---

[3]In Sec. 12.2 we will also consider scenarios of quintessence where the implied variation is non-monotonic.

Scenario 2

Scenario 2 favors a negative variation of α_X at BBN, and a negative variation may also fit the Mα results. However, the Reinhold μ measurement indicates a positive, but much smaller, variation. The Rμ results dominate the weighted average for l_4 due to their small error on $\Delta \ln \alpha_X$. The ratio $\Delta \ln \mu / \Delta \ln \alpha = 27$ makes this scenario unlikely to fit the reported signal of nonzero $\Delta \alpha$.

Scenario 3

In scenario 3 a positive variation of $\langle \phi \rangle / M_{GUT}$ is favored by BBN. The high ratio $\Delta \ln \mu / \Delta \ln \alpha \simeq -325$ makes the bounds obtained on a variation of μ strongly inconsistent with the claimed size of variation of α. The Reinhold et al. values again dominate the results for l_4.

Scenario 4

In this scenario, the ratio $\Delta \ln \mu / \Delta \ln \alpha = -22$ is again large and makes any observation of significant nonzero $\Delta \ln \alpha$ unlikely. Both the Mα and the Rμ measurements point in opposite direction to BBN; however the two spectroscopic observations are also inconsistent with each other, within the scenario. Again, the Rμ results dominate the determination of l_4 due to the small error.

Scenario 5, $\tilde{\gamma} = 42$

In this scenario the variation of α_X favored by BBN is positive ($l_6 = (1.7 \pm 1) \times 10^3$), however both nonzero variations from spectroscopic data Mα and Rμ require negative variations. With $\Delta \ln \mu / \Delta \ln \alpha = -6$ the spectroscopic measurements appear consistent with each other. Hence one would require some non-monotonic evolution to fit nonzero variations both at BBN and at moderate z. In Table 11.1 we have also evaluated l_6 using only the constraints given by D and ^{4}He (in brackets).

Scenario 6, $\tilde{\gamma} = 70$

As in the preceding scenario, BBN favors a positive variation in α_X, but Mα and Rμ favor negative. Again, Fig. 11.6 may suggest a non-monotonic evolution. Fitting to BBN including ^{7}Li we would obtain $l_6 = (1.2 \pm 0.6) \times 10^{-3}$; Tab. 11.1 also displays in brackets the value of l_6 obtained from D and ^{4}He bounds only.

Scenario 6, $\tilde{\gamma} = 25$

In this scenario, both BBN and the Mα signal favor a negative variation of α_X, whereas the Rμ observations point towards a positive variation. Following the argument of Wendt et al. [Wendt08], we substitute the Rμ value by the null constraint $|\Delta \mu / \mu| \leq 2.5 \times 10^{-5}$ [Wendt08] to obtain the bracketed value of l_4 in Table 11.1. In this scenario the evolution factors show a crossover from negligible variation at low redshift, to strong and monotonically increasing negative variation at $z \approx 2$.

11.3.4 Tension between the ^{7}Li problem and variation of μ

Measurements of the primordial ^{7}Li abundance show that the BBN abundance needs to decrease below the standard value to fit the observations, whereas the Reinhold μ measurement indicates μ to increase at $z \simeq 3$. We find that for all our unification scenarios the sign of the dependence on the fundamental parameter is the same for μ and ^{7}Li. Moreover, the coefficients of this dependence are nearly identical up to a common factor; hence the induced variations for μ and ^{7}Li point in the same direction, in contradiction to the tendency inferred from the observations. For example, for scenario 5 we find

$$\Delta \ln \mu = (23.2 - 0.65\tilde{\gamma})\Delta \ln \alpha_X,$$
$$\Delta \ln {}^7\text{Li} = (1692 - 49\tilde{\gamma})\Delta \ln \alpha_X. \qquad (11.46)$$

These expressions change sign at $\tilde{\gamma} = 35.7$ and 34.5, respectively. For a monotonic evolution, there is no possibility to have both a significant variation of μ and a variation of opposite sign in the ^{7}Li abundance. (In the regime $\tilde{\gamma} \approx 35$ there is no 2σ fit to BBN.) A similar result can be found for scenario 6 (including the SUSY partner mass dependence, which shows the same sort of degeneracy). Note that scenario 2 and 3 are just limiting cases of scenarios 5 and 6.

The main reason for this behavior is that variations of ^{7}Li and μ are dominated by the variations of $\hat{m}/\Lambda_{QCD}$ and m_e/Λ_{QCD}, respectively, with the same sign of prefactor. This degeneracy can be broken if m_e varies differently from the quark masses, a possibility that we do not consider in this thesis. For our scenarios with constant $\hat{m}/m_e$, the conflict between a monotonic time evolution and the μ- and ^{7}Li-observations is reflected in the opposite signs of l_4 and l_6.

This observational tension for monotonic behavior is clearly depicted in Fig. 11.8, where we plot simultaneously the averaged observational values of evolution factors $l_i/\ln(1 + z_i)$, normalized to $l_4/\ln(1 + z_4)$. For Scenario 6, $\tilde{\gamma} = 25$, we also display the result obtained by substituting the Wendt *et al.* value of μ variation for that of [Reinhold06]. The factor $\ln(1 + z_i)$ is introduced as a convenient normalization to avoid compressing the scale of variations excessively in recent epochs.[4] For the purpose of a quick inspection we have omitted the error bars, which are of course necessary for a quantitative interpretation.

---

[4] In quintessence-like theories, if the scalar field contributes a constant fraction of the total energy density of the Universe, as in so-called "tracker" models, the evolution of the field is typically also proportional to $\ln(1 + z)$. This is an additional motivation for our normalization.

11.3. EPOCHS AND EVOLUTION FACTORS

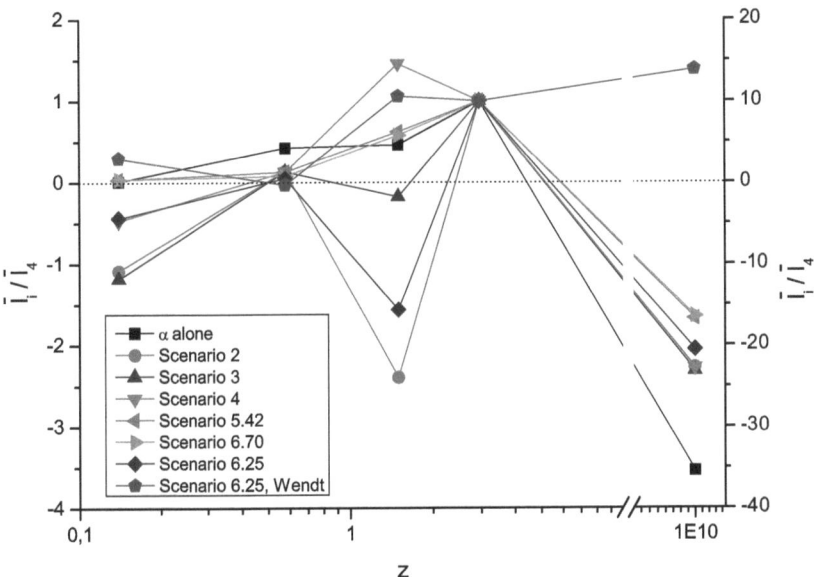

Figure 11.8: Normalized evolution factors $\bar{l}_i/\bar{l}_4$ for each scenario, where $\bar{l}_i \equiv l_i/\ln(1+z_i)$.

Chapter 12

Probing quintessence models

In this chapter we will describe how measurements of varying constants can be used to constrain models of quintessence under the assumption of grand unification. For that purpose we will use the measurements which we listed in Chap. 10 in the six unified scenarios studied in the preceding chapter to constrain the free parameters of the quintessence models described in Chap. 5. Our aim is to see to what extent such models can be consistent with the behavior of time variations that we have outlined.

12.1 Crossover quintessence

As we have seen in Sec. 5.3, crossover quintessence models yield a monotonic evolution of the cosmon and hence also a monotonic variation of constants, assuming a constant coupling δ to the fundamental varying parameter,

$$\Delta \ln \alpha_X(z) = \delta(\varphi(z) - \varphi(0)). \tag{12.1}$$

We discussed the viability of monotonic evolution in Sec. 11.3.3 where a first judgment can be made by inspection of Fig. 11.1 to 11.7 for the various unification scenarios, or by inspection of Table 11.1.

To allow us to easily compare with the observational results, we observe that in Table 11.1 the constraints for l_5 are considerably weaker than those for the remaining evolution factors. Furthermore, the one sigma range for l_4 is nonzero for all scenarios. Hence we shall compare observational and theoretical values for the ratios l_1/l_4, l_2/l_4, l_3/l_4 and l_6/l_4. We note the opposite sign of l_3 and l_4 for scenario 2, 3 and scenario 6 with $\tilde{\gamma} = 25$, which disfavors any monotonic evolution for these scenarios. The averaged observational values in each epoch are given in Table 12.1. Note that the coupling δ drops out of the ratios l_i/l_j. Considering these ratios allows us to probe quintessence directly without knowing the absolute size of the coupling, only assuming its (approximate) constancy over the relevant range of evolution. In view of its monotonic evolution, crossover quintessence cannot give negative ratios l_i/l_j. Hence it cannot be a good fit to the lithium abundance within the unification scenarios 2 to 6 that we consider. This reflects the tension between the [7]Li problem and a positive variation of μ discussed in Sec. 11.3.4.

For a simulation of the crossover quintessence model, we follow the procedure described in Sec. 5.3. By specifying the present densities of matter, radiation and dark energy and the model parameters w_{h0} and z_+, we can trace back the evolution

12.1. CROSSOVER QUINTESSENCE

Scenario	l_1/l_4	l_2/l_4	l_3/l_4	l_6/l_4
0	0.00 ± 0.01	0.13 ± 0.13	0.31 ± 0.18	-592 ± 4031
2	-0.10 ± 0.12	0.04 ± 0.04	-1.59 ± 0.88	-380 ± 228
3	-0.11 ± 0.13	0.04 ± 0.03	-0.12 ± 0.91	-387 ± 241
4	-0.04 ± 0.10	0.04 ± 0.03	0.97 ± 0.64	-381 ± 259
5, $\tilde{\gamma} = 42$	0.00 ± 0.03	0.04 ± 0.03	0.41 ± 0.19	-280 ± 191
without BBN	,,	,,	,,	-66 ± 165
6, $\tilde{\gamma} = 70$	0.00 ± 0.03	0.02 ± 0.02	0.39 ± 0.17	-275 ± 150
without BBN	,,	,,	,,	-69 ± 139
6, $\tilde{\gamma} = 25$	-0.04 ± 0.06	0.02 ± 0.04	-1.04 ± 0.49	-343 ± 142
with Wendt	0.03 ± 0.05	-0.01 ± 0.03	0.70 ± 0.52	231 ± 163

Table 12.1: Ratios of the evolution factors from observations.

w_{h0}	z_+	l_1/l_4	l_2/l_4	l_3/l_4	l_6/l_4
-0.95	3	0.12	0.38	0.67	26
-0.99	3	0.09	0.28	0.55	38
-0.9999	3	0.02	0.10	0.37	54
-0.95	7	0.15	0.47	0.80	13
-0.99	7	0.15	0.47	0.80	24
-0.9999	7	0.08	0.27	0.52	125
-0.999999	7	0.02	0.09	0.34	177

Table 12.2: Ratios of evolution factors from crossover quintessence. The l_i are evaluated by averaging over the variations evaluated at the same redshift as the data in each epochs (weighting by the number of absorption systems if appropriate).

of quintessence and the other components of our Universe. In Tab. 12.2 we display the ratios l_i/l_4 expected from crossover quintessence for various parameters w_{h0} and z_+, which can be compared with the observed ratios displayed in Tab. 12.1. Considering the lithium abundance to be affected by astrophysical systematics in scenarios 5 and 6 ($\tilde{\gamma} = 70$), or using the null result for $\Delta \ln \mu$ at intermediate redshift for scenario 6 ($\tilde{\gamma} = 25$ "with Wendt"), we find that some crossover quintessence models indeed yield the observed order of magnitude for the ratios l_i/l_4. We conclude that crossover quintessence could, in principle, reconcile a coupling variation of the claimed size in epochs 3 and 4 with the bounds from late cosmology, i.e. epochs 1 and 2. This is due to the "slowing down" of the cosmon evolution, as noted in [Wetterich03]. Values of the present equation of state w_{h0} quite close to -1 would be required, however. In other words, an observation of coupling variations would put strong bounds on the dynamics of the cosmon field[1] and provide for an independent source of information about the properties of dark energy.

Note that observational probes of dark energy would *not* give results for w_{h0} that coincide with the values that we take in our model. Such probes do not actually measure the present-day equation of state, rather they extrapolate w_0 from past epochs under some parameterization.

---

[1] The current WMAP5 bound on the equation of state parameter of quintessence is $w = -0.972 \pm 0.06$ [WMAP5], but the derivation assumes a constant w over a certain range of redshift, while our w_{h0} gives the equation of state at the present time.

12.2 Models with growing neutrinos and oscillating variation

In this section we investigate the two models of growing neutrinos which were laid out in Sec. 5.4. These models do not obey the proportionality of all coupling variations for all redshifts and do not show a monotonic evolution of the cosmon field. A systematic analysis of all such models seems difficult, hence we concentrate on the specific examples given in [Amendola08] and [Wetterich08].

The combined variation is different for each of the unified scenarios. For all unified parameters except $\langle\phi\rangle/M_{GUT}$ we still have a proportionality for the variations at all z, since the variations are proportional to φ. The variations due to the direct coupling (5.21) can be described by our method of evolution factors l_n, even though the l_n need not be strictly monotonic due to the oscillations in φ. However, for $\langle\phi\rangle/M_{GUT}$ we now have one variation linearly proportional to the variation of φ, Eq. (5.21) and an additional one with a nonlinear dependence, Eq. (5.30-5.31). A simple treatment with common evolution factors for all variations will no longer be applicable. Due to the additional φ-dependence of $\langle\phi\rangle/M_{GUT}$ we may have separate evolution factors for $\langle\phi\rangle/M_{GUT}$, different from the (common) evolution factors for the other couplings.

For example, the "linear contribution" (5.21) may dominate at BBN and induce a positive l_6 common for all couplings. In the range $z < 10$ the "non-linear contribution" (5.30) could be more important, leading to effectively negative $l_{3,4}$ for $\langle\phi\rangle/M_{GUT}$. (Such an effect could, in principle, relieve the tension between ^{7}Li and a positive μ-variation at high z which was explained in Sec. 11.3.4.) In practice, we calculate $\Delta\ln\alpha_X$ and $\Delta\ln\frac{\langle\phi\rangle}{M_{GUT}}$ at each epoch directly from the model, and extract the varying couplings and observables as explained in Chap. 11. Then we may search for a set of parameters δ, R_0 which minimizes the χ^2 for all measured variations.

12.2.1 The stopping growing neutrino model

The stopping growing neutrino model which was introduced in Sec. 5.4.1 has an oscillation in $\langle\phi\rangle$ that grows both in frequency and amplitude at late times as φ approaches its asymptotic value. Such oscillations must not be too strong as measurements between $z=2$ and today would measure a high rate of change. The oscillation may be made arbitrarily small by choosing small R_0. The restrictions from the low-z epochs are actually so strong that to a good approximation the non-linear contribution $\sim R_0$ can be neglected. However, the linear variation (5.21) is independent of R_0. It can be described by our method of evolution factors and yields for the set of parameters given in [Wetterich08] ($\varphi_t \approx 27.6$, $\alpha = 10$, $\epsilon = -0.05$) the ratios

$$\begin{aligned} l_1/l_4 &= 0.008, \\ l_2/l_4 &= 0.09, \\ l_3/l_4 &= 0.44, \\ l_6/l_4 &= 175. \end{aligned} \tag{12.2}$$

Comparing this with the numbers given in Table 12.1 shows that this model naturally yields evolution factors which are of the correct order of magnitude. We emphasize that no new parameter has been introduced for this purpose.

12.2.2 Global fit to the scaling growing neutrino model

Each growing neutrino model contains a few parameters that determine the cosmological evolution of the cosmon φ, and the Higgs v.e.v. $\langle\phi\rangle$. Two coupling parameters give respectively the relative strength of variation of α_X with φ, and the relative strength of the additional variation of the Higgs v.e.v. due to the varying triplet. For each example of cosmological evolution we may calculate the observables directly in terms of the two coupling parameters and make a global fit for their values. In performing the fit we take the 125 systems of the Murphy et al. α determination [Murphy03.2] within Epochs 3 and 4 (Eq. (10.12)) and further split them into 5 subsamples each with 25 absorption systems, since the data set extends over a wide range of redshift where there may be significant oscillations.

For the global fits, we take the scaling growing neutrino model (Sec. 5.4.2) with $\beta = -52$, $\alpha = 10$, $m_{\nu,0} = 2.3 \text{eV}$. With zero variation at all times (no degrees of freedom), we find $\chi^2 = 3.25$ including ^7Li at BBN, and $\chi^2 = 2.40$ neglecting ^7Li. The results of the best fits with varying couplings are given in Table 12.3.

Scenario	$\delta \times 10^4$	R_0	χ^2	$\Delta\chi^2$
2	-0.019	0.045	3.09	0.16
2 without Li7	-0.040	0	2.09	0.31
3	1.64	0	2.85	0.40
3 without Li7	1.55	0	2.03	0.37
4	3.79	0	2.85	0.40
4 without Li7	3.80	0	1.98	0.42
5.42	0.30	0	1.96	1.29
5.42 without Li7	0.24	0	1.87	0.53
6.70	0.20	0	1.93	1.32
6.70 without Li7	0.16	0	1.87	0.53
6.25	0.18	0.090	2.73	0.52
6.25 without Li7	-0.055	0.061	2.20	0.20

Table 12.3: Fitting parameters and minimal χ^2 values for the different unification scenarios for best fit to the scaling growing neutrino model [Amendola08]. The last column gives the increase in χ^2 produced when δ and R_0 are forced to vanish, i.e. for zero variation.

It turns out that χ^2 cannot be reduced by more than 1.3 in the fit including ^7Li and 0.53 in the fit neglecting ^7Li, which we do not consider as convincing evidence for coupling variations within this model. We have investigated some other choices of parameters for the cosmological evolution and also the stopping growing neutrino model, without a substantial change in the overall situation. In view of the unsettled status of the observational data it seems premature to make a systematic scan in parameter space. Our investigation demonstrates, however, how a clear positive signal for a coupling variation could restrict the parameter space for quintessence models.

Most of the additional variation of the Higgs v.e.v. occurs at later epochs, $z < 2$, thus recent observational bounds rule out any significant additional growth in $\langle\phi\rangle$. We considered fitting the observational values excluding BBN, as a function of the model parameters δ and R_0, and we find always that the value of R_0 at the minimum of χ^2 is unobservably small.

12.3 Tests of the weak equivalence principle

In Sec. 2.4 we have explained that variations of constants will influence the outcome of tests of the weak equivalence principle (WEP). Besides variations of couplings, the cosmon coupling to atoms also influences the outcome of tests of the weak equivalence principle, as it may also produce local gravitational effects which violate the WEP due to their interactions with matter. In Sec. 12.4 we use both aspects to set further constraints on a possible time variation of couplings. Test bodies with different composition have in general different couplings to the cosmon, and will hence experience different accelerations towards a common source. Usually, deviations from the universality of free fall are measured in terms of the Eötvös parameter η [Wetterich02.1, Dent06]

$$\eta^{b-c} \equiv \frac{2|a_b - a_c|}{|a_b + a_c|} , \qquad (12.3)$$

where $a_{b,c}$ are the accelerations towards the source of the two test masses. The experiment setting the currently tightest limits on η [Schlamminger07] has the result

$$\eta = (0.3 \pm 1.8) \times 10^{-13} \qquad (12.4)$$

for test bodies of Be ($A = 9$, $Z = 4$) and Ti ($A = 48$, $Z = 22$) composition, where the gravitational source is taken to be the Earth.

In contrast to the direct observations of time varying couplings, tests of the universality of free fall do not determine directly the values of fundamental 'constants' or their possible variations. However, given our basic assumption of a slow time variation, driven by a light scalar degree of freedom, the current limits on composition-dependent long range forces put bounds on the scalar couplings to different constituents of matter. In our language, they measure or constrain the coefficients β_k at $z = 0$, which relate the evolution factors and the changes in the cosmon field (see Eq. (11.42)),

$$\Delta \ln G_k = d_k l = \beta_k \Delta \varphi(z_n) . \qquad (12.5)$$

These constraints then imply bounds on the time variation of constants: Differentiating with respect to time, Eq. (12.5) becomes

$$\frac{\dot{G}_k}{G_k} = \beta_k \dot{\varphi} . \qquad (12.6)$$

Applying the conservative bound [Dent06] $\dot{\varphi}/H_0 \leq 0.7$, we derive

$$\dot{\varphi} \leq \dot{\varphi}_{\max} \simeq 5 \times 10^{-11} \text{y}^{-1} . \qquad (12.7)$$

β_k can be derived as a function of η, utilizing a model of nuclear masses and binding energies and using GUT relations to reduce the number of free parameters (see [DSW08.2] for details on this treatment). It turns out that WEP violation places significant bounds on the present-day values of scalar couplings, as will be shown in the next section.

The differential acceleration η for two bodies with equal mass but different composition and therefore different "cosmon charge" is related to the present time variation of couplings and cosmological parameters via [DSW08.2]

$$\eta \simeq 3.8 \times 10^{-12} \left(\frac{\dot{\alpha}/\alpha}{10^{-15}\text{y}^{-1}} \right)^2 \frac{F}{\Omega_h(1+w_h)} . \qquad (12.8)$$

Scenario	2	3	4	5, $\tilde{\gamma}=42$	6, $\tilde{\gamma}=70$	6, $\tilde{\gamma}=25$
F (Be-Ti)	95	-9000	-165	-25	-26	41

Table 12.4: Values of F for a WEP experiment using Be-Ti masses for the unified scenarios studied in this thesis.

Here $\dot{\alpha}/\alpha$ is the present relative variation of the fine structure constant in units of year^{-1}, w_h is the present dark energy equation of state and $\Omega_h \approx 0.73$ the present dark energy fraction. The "unification factor" F encodes the information on the specific relations between variations in fundamental parameters as implied by the different GUT scenarios, and the composition of the test bodies. The factor F is displayed in Tab. 12.4 for our different unified scenarios and for the Be-Ti test masses of [Schlamminger07]. For typical test mass compositions we find $1 \leq F \leq$ few $\times 10^2$. The very large value of F for Scenario 3 reflects an accidental cancellation of terms which we do not consider to be typical.

Once F is fixed, the relation (12.8) allows for a direct comparison between the sensitivity of measurements of η versus the measurements of $\dot{\alpha}/\alpha$ from laboratory experiments, or bounds from recent cosmological history, such as from the Oklo natural reactor or the isotopic composition of meteorites.

12.4 Bounds on present-day variation

Within our theoretical framework there exist three distinct ways to bound or measure the present-day rate of variation of fundamental parameters. The first is a direct measurement, of the type probed by atomic clock experiments (see Sec. 10.7). If one or more nonzero variations are found in this way, bounds on unified models may immediately be set. The second method is by combining information on the size of scalar field couplings from WEP tests (Section 12.3) with a cosmological upper bound on the kinetic energy of scalar fields [Wetterich02.1, Dent06]. Such bounds on scalar couplings will depend on the choice of unified model and in general will be independent of those derived from atomic clocks. Thirdly, under the assumption of a monotonic variation (that also does not significantly accelerate with time), we may convert any "historic" bound on the net variation of a fundamental parameter since a given epoch into a bound on the present rate of variation:

$$|\dot{G}_k| \leq (t_0 - t_n)^{-1} |G_k(t_0) - G_k(t_n)| \equiv \frac{|\Delta G_k|}{\Delta t}, \quad t_n < t_0. \quad (12.9)$$

Here t_0 denotes the present, and any nonzero rate of variation should have the appropriate sign, i.e. $\dot{G}_k$ has the opposite sign to ΔG_k referring to some past epoch.

For any given unified model of time variations, the three bounds on present-day evolution will have different sensitivity. Therefore if one method gives a nonzero variation we would (in some cases) be able to distinguish between models due to the fact that the other bounds are still consistent with zero. To give a simple example, the direct detection of a nonzero time variation in atomic clocks near the present upper bound would immediately rule out a large class of models that cannot account for such a variation without leading to WEP violation above current bounds; and would also rule out models in which such nonzero variations extrapolated to past epochs $t_n < t_0$ would exceed observational bounds.

However, this chain of inference does not function equally well in all directions. A nonzero finding of differential acceleration violating the WEP would indicate nontrivial scalar couplings, but need not imply nonzero time variation since the rate of change of the scalar is not bounded from below. Also, a nonzero variation at some past epoch t_n would not necessarily imply a lower bound to the present-day rate of variation or size of scalar couplings, since the variation could have slowed substantially since then (either due to nonlinear scalar evolution or a nonlinear coupling function). Only with the assumption of a reasonably smooth and monotonic variation of the scalar field and its coupling functions, one can find, for any given unified model, where the first signals of present-day or recent variation are expected to appear.

At present these methods give null results up to redshifts of about 0.8, but if a nonzero time variation exists, we can determine for each unified scenario which observational method is most sensitive. Thus if a nonzero signal of late time variation arises it may be used to distinguish between models. We assume for this purpose an approximately linear variation over recent cosmological times, thus measurements of absolute variation at nonzero redshift z imply time derivatives

$$\frac{d \ln X}{dt} \simeq \frac{\Delta \ln X(z)}{t_0 - t(z)}. \tag{12.10}$$

Here X is the fundamental varying parameter: we consider first $X \equiv \alpha$, if only the fine structure varies; in scenario 1 $X \equiv G_N m_N^2$, in scenarios 2, 5 and 6 $X \equiv \alpha_X$ and in scenarios 3 and 4 $X \equiv \langle \phi \rangle / M_{GUT}$. Then Table 12.5 gives the precision of bounds on time derivatives for the unified scenarios we consider, except scenario 1 (varying G_N) which is probed by a quite different set of measurements. As explained in Sec. 10.4, we take the Oklo bound as applying directly to the variation of α, and increase its uncertainty by a factor 3 to account for possible cancellations when other parameters also vary. We present the recent Rosenband et al. [Rosenband08] Al/Hg ion clock bound separately to illustrate to what extent it improves over previous atomic clock results.

Extending this method beyond $z \approx 0.5$ becomes questionable. One could use linearity in $\ln(1 + z)$ instead of t, but even this improvement may lead to unreliable extrapolations for models with a particular dynamics of the scalar field, as crossover quintessence.

12.4. BOUNDS ON PRESENT-DAY VARIATION

Scenario	X	Clocks	Al/Hg	WEP	Oklo α	Meteorite	Astro
α only	α	0.13 (α)	0.023	6.2	0.033	0.32	0.44 (y)
2	α_X	0.074 (μ)	0.027	0.007	0.12	0.015	0.006 (NH_3)
2S	α_X	0.12 (μ)	0.044	0.012	0.19	0.026	0.010 (NH_3)
3	$\langle\phi\rangle/M_X$	2.6 (μ)	12.4	0.33	54	0.53	0.22 (NH_3)
4	$\langle\phi\rangle/M_X$	6.2 (μ)	1.78	0.35	7.7	1.2	0.51 (NH_3)
5, $\tilde{\gamma} = 42$	α_X	0.32 (α)	0.024	0.013	0.11	0.069	0.035 (NH_3)
6, $\tilde{\gamma} = 70$	α_X	0.21 (α)	0.016	0.008	0.070	0.049	0.025 (NH_3)
6, $\tilde{\gamma} = 25$	α_X	0.25 (μ)	0.027	0.011	0.12	0.056	0.021 (NH_3)

Table 12.5: Competing bounds on recent ($z \leq 0.8$) time variations in unified scenarios. For each scenario we give 1σ uncertainties of bounds on $d(\ln X)/dt$ in units 10^{-15}y^{-1}, where X is the appropriate fundamental parameter. For the Oklo bound we multiply the uncertainty by a factor 3 except when only α varies. The column "Clocks" indicates whether α or μ gives the stronger bound; the recent Al/Hg limit [Rosenband08] is given a separate column. The column "Astro" indicates which measurements of astrophysical spectra are currently most sensitive in each scenario.

Chapter 13

Conclusion and outlook

We have developed a systematic method to relate variations of fundamental parameters of particle physics to the primordial isotope abundances produced by BBN. The main advantage of the method, which is laid out in part two of this thesis, is that we are able to vary every parameter independently, both at the level of fundamental Standard Model parameters and of nuclear physics parameters, thus we are not dependent on any particular theoretical model which enforces particular relations between the variations.

We follow a two step approach, first extracting the nuclear parameter dependence (without major theoretical uncertainties) and in a second step relating this to variations of fundamental Standard Model parameters. We define two linear response matrices, where the first, C, encodes the change in predicted abundances produced by small variations away from the current values of nuclear physics parameters which enter the BBN integration code. These parameters comprise the gravitational constant, fine structure constant, neutron lifetime, electron, proton and neutron masses, and binding energies of $A \leq 7$ nuclei. The dependences of nuclear reaction rates on these parameters are also implemented insofar as they are calculated within some effective theory. One notable result is that the ^{7}Li abundance depends heavily on the binding energies of ^{3}He, ^{4}He and ^{7}Be.

We also investigated possible further effects of variations in nuclear reaction rates on predicted abundances by varying each rate (*i. e.* thermal integrated cross-section $\langle \sigma v \rangle$) separately by a temperature-independent factor. We find that the ^{4}He abundance is insensitive to nuclear rates, and only eight reactions could lead to significant variation of the D, ^{3}He or ^{7}Li abundances.

The second response matrix, F, relates variations in nuclear parameters to the fundamental parameters of particle physics, comprising the gravitational constant, fine structure constant, Higgs vacuum expectation value, electron mass, and the light (up and down) quark masses. At this point theoretical uncertainty enters into the relation between quark masses and nuclear binding energies. We parameterize the dependence of binding energies on the pion mass (and hence on light quark masses) by the deuteron binding, which has been treated by a systematic expansion in effective field theory.

The resulting fundamental response matrix $R = CF$ allows us, first, to bound the variations of the six fundamental couplings individually, some bounds being at the percent level. Secondly, studying three exemplary unified scenarios, we can also bound correlated variations affecting many couplings at once. We find that one scenario allows us to fit observed D, ^{4}He and ^{7}Li abundances within 2σ bounds, given a variation $\Delta\alpha/\alpha = -2 \times 10^{-4}$ away from the present value; another fits these observational abundances within almost 1σ bounds, given a variation $\Delta\alpha/\alpha = 4 \times 10^{-4}$. For this analysis we have left the linear matrix approach and make use of our BBN code which allows us to derive the full nonlinear dependence.

In a next step (part three of the thesis), we combine our findings for BBN with further measurements of the variation of fundamental constants from BBN to today. Within grand unified theories, this set of different varying parameters can be consistently reduced to a variation of a few "unification parameters", namely the unification scale M_{GUT}/M_P, gauge coupling α_X, the Fermi scale $\langle\phi\rangle/M_{GUT}$ and SUSY-breaking masses $\tilde{m}/M_{GUT}$. We define various GUT-scenarios for varying couplings by the assumption of proportionality of fractional variations of the unification parameters.

Assuming that couplings really vary, this is a way of excluding such GUT scenarios by demanding consistency of the implied variations. The assumption of proportionality permits us to project all observations into constraints on a common evolution factor $l(z)$ for each scenario. We show that different GUT scenarios yield different time evolutions of $l(z)$ assuming that certain claimed measurements of varying constants are correct. We confirm that "simple" models which have only one fundamental parameter varying (α_X or $\langle\phi\rangle/M_{GUT}$) result in inconsistent variations. However, combined variations of these two parameters, as described in scenarios 5 and 6, lead to results more consistent with the possible quintessence-induced time variations of fundamental couplings. For instance, some models of crossover quintessence and models of growing neutrinos naturally yield ratios of evolution factors which are of the same size as those derived from measurements of varying couplings.

Still, we have not found a scenario with a monotonic time evolution $l(z)$ that makes the two main signals or hints of variation ([Murphy03.2], [Reinhold06]) and BBN mutually consistent. A monotonic evolution requires either to discount one of the "signals" by substantially increasing its uncertainty, or to alter our assumptions by including additional time variation of some Yukawa couplings.

In a last step we demonstrate how a clear observation of time variation of fundamental couplings would not only strongly disfavor a constant dark energy, but also put important constraints on the time evolution of a dynamical dark energy or quintessence. We have shown that for a given unified scenario, the bounds on the time variation of various couplings in different cosmological epochs can strongly restrict the possible time evolution of the cosmon field and put very strong bounds on late-time dynamics of the cosmon field. Hence, once at least one irrefutable observation of some coupling variation at some redshift becomes available, our method provides a powerful tool for testing extensions of the Standard Model or, vice versa, allows us to control consistency of claimed variations under the demand of unification.

We have demonstrated this by an analysis that implicitly assumes a nonzero variation, considering both general features and specific quintessence models. However, we are aware that the actual values for the evolution factors $l(z)$ from this analysis may be premature, since the observational situation is unclear and on moving grounds. For example, taking the recent reanalysis of the variation of the proton to electron mass ratio μ in Ref. [King08] instead of the results in Ref. [Reinhold06] used in this thesis, would strongly influence the values of the evolution factors. We have demonstrated this in a somewhat different way by investigating the change in the evolution factors if some claimed observations of varying couplings are omitted.

Outlook

Progress in the field of BBN requires both observational and theoretical improvements. Both statistical and systematic errors in abundance measurements could be improved, for example observations to better determine the nature of systems where ^{4}He is measured [Steigman05], or stellar modeling to test possible solutions of the ^{7}Li problem. On the theoretical side the relation between quark masses and nuclear physics remains unclear beyond the level of the two-nucleon system: the largest uncertainty in our BBN bounds arises from the poorly known dependence of the binding energies on the fundamental couplings. BBN is already the most powerful probe of fundamental "constants" in the early Universe, and precision bounds may well be obtained, given continued efforts in observation and theory, to rule out or confirm the presence of a cosmological variation.

Our investigation has further shown how the variations of different couplings in the Standard Model may be compared. If the observational situation becomes clearer and at least one nonzero time variation is established, such methods may be used for new tests of the idea of grand unification and, even further, models of quintessence. The presented method could then easily be applied to constrain both theories beyond the Standard Model like grand unification and theories beyond the concordance model of cosmology like quintessence. Hence, it can establish relations between two a priori distinct theories, namely high-energy physics and general relativity, a capability which is so far dominated by theories with radically new concepts of physics like string theory.

Acknowledgements

First of all I would like to thank Christof Wetterich, the supervisor of this thesis, for giving me the chance to work on the highly interesting questions of varying constants, cosmology and quintessence. His offer allowed me to stay at the beautiful Institute for Theoretical Physics in Heidelberg, a place which I will always remember.

Furthermore, I would like to thank Thomas Dent for almost three years of deep and successful cooperation. His valuable hints and explanations opened the door to new insights and approaches which I would not have found that easily without him.

The members of the Institute and especially the "Dachzimmer" deserve gratitude for the wonderful atmosphere. Thanks to Georg Robbers and Thomas Dent for proofreading this thesis.

The final thanks go to the *Studienstiftung des Deutschen Volkes*. I am greatly indebted to them for supporting me during my whole studies and PhD, which made important things possible in my life.

Appendix A

Conventions

A.1 Symbols and abbreviations

Symbol	Explanation
α, α_{em}	fine structure constant
α_S	strong coupling constant
α_X	grand unified coupling constant
B_i	binding energy of nucleus i
δ_q	light quark mass difference
η	baryon-to-photon ratio; Eötvös parameter
G_N	Newton's constant
Λ_{QCD}	QCD scale
$\hat{m}$	average light quark mass
m_d	d-quark mass
m_e	electron mass
m_n	neutron mass
m_π	average pion mass
m_p	proton mass
m_s	strange quark mass
m_u	u-quark mass
m_N	average nucleon mass, $m_N = \frac{1}{2}(m_n + m_p)$
M_{GUT}	GUT scale
M_P	Planck mass
$\bar{M}_P$	reduced Planck mass
M_X	GUT scale ($\equiv M_{GUT}$)
μ	proton to electron mass ratio, $\mu := m_p/m_e$
Ω_b	baryon fraction
Q_N	neutron proton mass difference, $Q_N := m_n - m_p$

Table A.1: List of symbols

Abbreviation	Explanation
BAO	baryon acoustic oscillations
BBN	Big Bang Nucleosynthesis
EFT	effective field theory
CMB	cosmic microwave background
DOF	degree of freedom
MSSM	minimal supersymmetric standard model
QCD	quantum chromodynamics
QED	quantum electrodynamics
SBBN	standard big bang nucleosynthesis
SM	Standard Model (of particle physics)
SN	supernovae
v.e.v.	vacuum expectation value
WEP	weak equivalence principle

Table A.2: List of abbreviations

List of Tables

2.1	The fundamental constants of nature	5
3.1	Equation-of-state parameters for different types of cosmological components	10
3.2	Parameters describing our Universe	15
4.1	The MSSM particle spectrum	21
4.2	Renormalization group coefficients for the strong interaction	22
4.3	Renormalization group coefficients for the electromagnetic interaction	22
6.1	Leading dependence of abundances on thermal averaged cross-sections	43
6.2	Current observational and theoretical primordial abundances	53
7.1	Response matrix C, dependence of abundances on nuclear parameters.	56
8.1	Response matrix F, dependence of nuclear on fundamental parameters	63
8.2	Response matrix R, dependence of abundances on fundamental parameters	63
8.3	Dependence of abundances on fundamental parameters found by [MSW04]	64
8.4	Dependence of abundances on fundamental parameters from [Chamoun05] and [Landau04]	64
8.5	Dependence of abundances on α from [Bergstrom99] and [Nollett02] .	64
9.1	Allowed individual variations of fundamental couplings	65
10.1	Observational 1σ bounds on variations	81
11.1	Redshifts and evolution factors for each epoch	94
12.1	Ratios of the evolution factors from observations.	99
12.2	Ratios of evolution factors from crossover quintessence.	99
12.3	Fit to the scaling growing neutrino model	101
12.4	Values of F for the unified scenarios studied in this thesis	103
12.5	Competing bounds on recent time variations in unified scenarios . . .	105
A.1	List of symbols	109
A.2	List of abbreviations	110

List of Figures

3.1 Evolution of the components in a ΛCDM Universe 12
3.2 History of our Universe from Particle Data Group 2000 13
3.3 Today's energy content of our Universe 15

4.1 Running of coupling constants in the Standard Model and in SUSY . . 18
4.2 Unification of the four forces in the string theory picture 19

5.1 Equation of state of quintessence in the crossover quintessence model . 28
5.2 Energy components of our Universe in the crossover quintessence model 29
5.3 Dimensionless cosmon field φ in the crossover quintessence model . . . 29
5.4 Evolution of the cosmon field in the stopping growing neutrino model 32
5.5 Additional variation of the Higgs v.e.v. in the stopping growing neutrino model 32
5.6 Evolution of the cosmon field in the scaling growing neutrino model . 33
5.7 Additional variation of the Higgs v.e.v. in the scaling growing neutrino model 33

6.1 Network of main reactions responsible for primordial nucleosynthesis . 37
6.2 Element abundances as a function of the temperature of the universe . 38
6.3 ^{4}He abundances versus oxygen abundance 50
6.4 Deuterium versus neutral column density N_{HI} 51
6.5 ^{3}He abundances versus oxygen 51
6.6 ^{7}Li abundances versus metallicity 52

9.1 Variation of primordial abundances with α in three GUT scenarios . . 68
9.2 Variation of primordial abundances with α in three GUT scenarios including nonlinear effects 70

11.1 Variations for varying α alone 86
11.2 Variations for scenario 2 86
11.3 Variations for scenario 3 89
11.4 Variations for scenario 4 89
11.5 Variations for scenario 5, $\tilde{\gamma} = 42$ 90
11.6 Variations for scenario 6, $\tilde{\gamma} = 70$ 90
11.7 Variations for scenario 6, $\tilde{\gamma} = 25$ 91
11.8 Normalized evolution factors $\bar{l}_i/\bar{l}_4$ for each scenario 97

Bibliography

[Amaldi91]   U. Amaldi, W. de Boer, P. H. Frampton, H. Furstenau, and J. T. Liu. Consistency checks of grand unified theories. *Phys. Lett.*, B281:374–383, 1992.

[Amendola08] L. Amendola, M. Baldi, and C. Wetterich. Growing Matter. *Phys. Rev.*, D78:023015, 2008.

[Aprahamian05] A. Aprahamian, K. Langanke, and M. Wiescher. Nuclear structure aspects in nuclear astrophysics. *Progress in Part. and Nucl. Phys.*, 54:535, 2005.

[Asplund05]   M. Asplund, D. L. Lambert, P. E. Nissen, F. Primas, and V. V. Smith. Lithium isotopic abundances in metal-poor halo stars. *Astrophys. J.*, 644:229–259, 2006.

[Bania02]   T. M. Bania, R. T. Rood, and D. S. Balser. The cosmological density of baryons from observations of ^{3}He$^+$ in the Milky Way. *Nature*, 415:54–57, 2002.

[Banks01]   T. Banks, M. Dine, and M. R. Douglas. Time-varying alpha and particle physics. *Phys. Rev. Lett.*, 88:131301, 2002.

[Bartelmann06] M. Bartelmann. Observing the Big Bang. 2006. Lecture Notes.

[BeaneSavage02] S. R. Beane and M. J. Savage. The quark mass dependence of two-nucleon systems. *Nucl. Phys.*, A717:91–103, 2003.

[Berger06]   C. Berger. Elementary particle physics: From the foundations to the modern experiments. Berlin, Germany: Springer, 2006.

[Bergstrom99] L. Bergstrom, S. Iguri, and H. Rubinstein. Constraints on the variation of the fine structure constant from big bang nucleosynthesis. *Phys. Rev.*, D60:045005, 1999.

[Blatt08]   S. Blatt et al. New Limits on Coupling of Fundamental Constants to Gravity Using ^{87}Sr Optical Lattice Clocks. *Phys. Rev. Lett.*, 100:140801, 2008.

[Bonifacio06] P. Bonifacio et al. First stars VII. Lithium in extremely metal poor dwarfs. 2006. [arXiv:astro-ph/0610245].

[Borasoy96]   B. Borasoy and U.-G. Meissner. Chiral expansion of baryon masses and sigma-terms. *Annals Phys.*, 254:192–232, 1997.

[Brans05] C. H. Brans. The roots of scalar-tensor theory: An approximate history. 2005. [arXiv:gr-qc/0506063].

[BransDicke61] C. Brans and R. H. Dicke. Mach's principle and a relativistic theory of gravitation. *Phys. Rev.*, 124:925–935, 1961.

[Buchert07] T. Buchert. Dark Energy from Structure - A Status Report. *Gen. Rel. Grav.*, 40:467–527, 2008.

[Carter74] B. Carter. Large number coincidences and the anthropic principle in cosmology. *IAU Symp.*, 63:291, 1974.

[Chamoun05] N. Chamoun, S. J. Landau, M. E. Mosquera, and H. Vucetich. Helium and deuterium abundances as a test for the time variation of the fine structure constant and the Higgs vacuum expectation value. *J. Phys.*, G34:163–176, 2007.

[Chan07] K.-C. Chan and M. C. Chu. Constraining the Variation of G by Cosmic Microwave Background Anisotropies. *Phys. Rev.*, D75:083521, 2007.

[Chandrasekhar37] S. Chandrasekhar. The Cosmological Constants. *Nature*, 139:757–758, 1937.

[Chanfray94] G. Chanfray, M. Ericson, and M. Kirchbach. Pion decay constant and the gell-mann-oakes-renner relation in nuclei. *Mod. Phys. Lett.*, A9:279–287, 1994.

[Charbonnel05] C. Charbonnel and F. Primas. The Lithium Content of the Galactic Halo Stars. 2005. [arXiv:astro-ph/0505247].

[Chen98] J.-W. Chen, H. W. Griesshammer, M. J. Savage, and R. P. Springer. The polarizability of the deuteron. *Nucl. Phys.*, A644:221–234, 1998.

[ChenRupak99] J.-W. Chen, G. Rupak, and M. J. Savage. Nucleon nucleon effective field theory without pions. *Nucl. Phys.*, A653:386–412, 1999.

[Chen99] J.-W. Chen and M. J. Savage. n+p → d+γ for big-bang nucleosynthesis. *Phys. Rev.*, C60:065205, 1999.

[ChristyDuck61] R.F. Christy and I. Duck. γ rays from an extranuclear direct capture process. *Nucl. Phys.*, 24:89, 1961.

[Coc06] A. Coc, N. J. Nunes, K. A. Olive, J.-P. Uzan, and E. Vangioni. Coupled Variations of Fundamental Couplings and Primordial Nucleosynthesis. *Phys. Rev.*, D76:023511, 2007.

[Copeland06] E. J. Copeland, M. Sami, and S. Tsujikawa. Dynamics of dark energy. *Int. J. Mod. Phys.*, D15:1753–1936, 2006.

[Damour07] T. Damour and J. F. Donoghue. Constraints on the variability of quark masses from nuclear binding. 2007. [arXiv:0712.2968].

BIBLIOGRAPHY

[Damour88]   T. Damour, G. W. Gibbons, and J. H. Taylor. Limits on the variability of g using binary-pulsar data. *Phys. Rev. Lett.*, 61(10):1151–1154, 1988.

[Dent03]   T. Dent. Varying alpha, thresholds and extra dimensions. 2003. [arXiv:hep-ph/0305026].

[Dent06]   T. Dent. Composition-dependent long range forces from varying m(p)/m(e). *JCAP*, 0701:013, 2007.

[Dent08]   T. Dent. Fundamental constants and their variability in theories of High Energy Physics. 2008. [arXiv:0802.1725].

[Dirac37]   P. A. M. Dirac. The Cosmological constants. *Nature*, 139:323, 1937.

[Dirac38]   P. A. M. Dirac. New basis for cosmology. *Proc. Roy. Soc. Lond.*, A165:199–208, 1938.

[Dirac79]   P. A. M. Dirac. The large numbers hypothesis and the Einstein theory of gravitation. *Proc. Roy. Soc. Lond.*, A365:19, 1979.

[Donoghue06]   J. F. Donoghue. The Nuclear Central Force in the Chiral Limit. *Phys. Rev.*, C74:024002, 2006.

[Doran07]   M. Doran, G. Robbers, and C. Wetterich. Impact of three years of data from the Wilkinson Microwave Anisotropy Probe on cosmological models with dynamical dark energy. *Phys. Rev.*, D75:023003, 2007.

[DST06]   M. Doran, S. Stern, and E. Thommes. Baryon Acoustic Oscillations and Dynamical Dark Energy. *JCAP*, 0704:015, 2007.

[DSW07]   T. Dent, S. Stern, and C. Wetterich. Primordial nucleosynthesis as a probe of fundamental physics parameters. *Phys. Rev.*, D76:063513, 2007.

[DSW08.1]   T. Dent, S. Stern, and C. Wetterich. Unifying cosmological and recent time variations of fundamental couplings. *Phys. Rev.*, D78:103518, 2008.

[DSW08.2]   T. Dent, S. Stern, and C. Wetterich. Time variation of fundamental couplings and dynamical dark energy. 2008. [arXiv:0809.4628], to appear in JCAP.

[Einstein17]   A. Einstein. Cosmological considerations in the general theory of relativity. *Sitzungsber. Preuss. Akad. Wiss. Berlin (Math. Phys.)*, 1917:142–152, 1917.

[Epelbaum02]   E. Epelbaum, U.-G. Meissner, and W. Gloeckle. Nuclear forces in the chiral limit. *Nucl. Phys.*, A714:535–574, 2003.

[Esmailzadeh91]   R. Esmailzadeh, G. D. Starkman, and S. Dimopoulos. Primordial nucleosynthesis without a computer. *Astrophys. J.*, 378:504–518, 1991.

[Feynman39]   R. P. Feynman. Forces in molecules. *Phys. Rev.*, 56:340–343, 1939.

[Fiorentini98]   G. Fiorentini, E. Lisi, S. Sarkar, and F. L. Villante. Quantifying uncertainties in primordial nucleosynthesis without Monte Carlo simulations. *Phys. Rev.*, D58:063506, 1998.

[FlambaumNH3]   V. V. Flambaum and M. G. Kozlov. Enhanced sensitivity to time-variation of m(p)/m(e) in the inversion spectrum of ammonia. 2007. [arXiv:0704.2301].

[Flambaum02]   V. V. Flambaum and E. V. Shuryak. Dependence of hadronic properties on quark masses and constraints on their cosmological variation. *Phys. Rev.*, D67:083507, 2003.

[Flambaum07]   V. V. Flambaum and R. B. Wiringa. Dependence of nuclear binding on hadronic mass variation. *Phys. Rev.*, C76:054002, 2007.

[Fornengo97]   N. Fornengo, C. W. Kim, and J. Song. Finite temperature effects on the neutrino decoupling in the early universe. *Phys. Rev.*, D56:5123–5134, 1997.

[Fortier07]   T. M. Fortier et al. Precision atomic spectroscopy for improved limits on variation of the fine structure constant and local position invariance. *Phys. Rev. Lett.*, 98:070801, 2007.

[Fowler67]   W. A. Fowler, G. R. Caughlan, and B. A. Zimmerman. Thermonuclear reaction rates. *Ann. Rev. Astron. Astrophys.*, 5:525–570, 1967.

[Fowler75]   W. A. Fowler, G. R. Caughlan, and B. A. Zimmerman. Thermonuclear reaction rates. 2. *Ann. Rev. Astron. Astrophys.*, 13:69–112, 1975.

[FowlerHoyle64]   W. A. Fowler and F. Hoyle. Neutrino processes and pair formation in massive stars and supernovae. *Astrophys. J. Suppl.*, 9:201–319, 1964.

[Fujii03]   Y. Fujii and A. Iwamoto. Re/Os constraint on the time-variability of the fine- structure constant. *Phys. Rev. Lett.*, 91:261101, 2003.

[Fujii07]   Y. Fujii. Possible time-variability of the fine-structure constant expected from the accelerating universe. *Phys. Lett.*, B660:87–92, 2008.

[Gasser75]   J. Gasser and H. Leutwyler. Implications of scaling for the proton - neutron mass - difference. *Nucl. Phys.*, B94:269, 1975.

[Gasser82]   J. Gasser and H. Leutwyler. Quark masses. *Phys. Rept.*, 87:77–169, 1982.

[Gegelia98]   J. Gegelia. Are pions perturbative in effective field theory? 1998. [arXiv:nucl-th/9806028].

[Gell-Mann68] M. Gell-Mann, R. J. Oakes, and B. Renner. Behavior of current divergences under su(3) x su(3). *Phys. Rev.*, 175:2195–2199, 1968.

[Glashow70] S. L. Glashow, J. Iliopoulos, and L. Maiani. Weak Interactions with Lepton-Hadron Symmetry. *Phys. Rev.*, D2:1285–1292, 1970.

[Guenther98] D.B. Guenther, L.M. Krauss, and P. Demarque. Testing the constancy of the gravitational constant using helioseismology. *Astrophys. J.*, 498:871–876, 1998.

[HalzenMartin] F. Halzen and A. D. Martin. Quarks and leptons: an introductory course in modern particle physics. 1984. New York, Usa: Wiley (1984).

[Hebecker00] A. Hebecker and C. Wetterich. Natural quintessence? *Phys. Lett.*, B497:281–288, 2001.

[Hellings83] R. W. Hellings, P. J. Adams, J. D. Anderson, M. S. Keesey, E. L. Lau, E. M. Standish, V. M. Canuto, and I. Goldman. Experimental test of the variability of g using viking lander ranging data. *Phys. Rev. Lett.*, 51(18):1609–1612, 1983.

[Hellmann37] H. Hellmann. Einfuehrung in die quantenchemie. 1937. Deuticke, Leipzig, 1937.

[Hemmert03] T. R. Hemmert, M. Procura, and W. Weise. Quark mass dependence of the nucleon axial-vector coupling constant. *Phys. Rev.*, D68:075009, 2003.

[HinshawWMAP5] G. Hinshaw et al. Five-Year Wilkinson Microwave Anisotropy Probe (WMAP) Observations:Data Processing, Sky Maps, & Basic Results. 2008. [arXiv:0803.0732].

[Kanekar05] N. Kanekar et al. Constraints on changes in fundamental constants from a cosmologically distant OH absorber / emitter. *Phys. Rev. Lett.*, 95:261301, 2005.

[Kawano88] L. Kawano. Let's go: Early universe. guide to primordial nucleosynthesis programming. 1988. FERMILAB-PUB-88-034-A.

[Kawano92] L. Kawano. Let's go: Early universe. 2. primordial nucleosynthesis: The computer way. 1992. FERMILAB-PUB-92-004-A.

[Kiener6Li] J. Kiener et al. Measurements of the Coulomb dissociation cross section of 156 MeV Li-6 projectiles at extremely low relative fragment energies of astrophysical interest. *Phys. Rev.*, C44:2195–2208, 1991.

[King08] J. A. King, J. K. Webb, M. T. Murphy, and R. F. Carswell. Stringent null constraint on cosmological evolution of the proton-to-electron mass ratio. 2008. [arXiv:0807.4366].

[Kirkman03] D. Kirkman, D. Tytler, N. Suzuki, J. M. O'Meara, and D. Lubin. The cosmological baryon density from the deuterium to hydrogen ratio towards QSO absorption systems: D/H towards Q1243+3047. *Astrophys. J. Suppl.*, 149:1, 2003.

[Korn06]   A. J. Korn et al. A probable stellar solution to the cosmological lithium discrepancy. *Nature*, 442:657–659, 2006.

[Landau04]   S. J. Landau, M. E. Mosquera, and H. Vucetich. Primordial nucleosynthesis with varying fundamental constants: A semi-analytical approach. *Astrophys. J.*, 637:38–52, 2006.

[Langacker01] P. Langacker, G. Segre, and M. J. Strassler. Implications of gauge unification for time variation of the fine structure constant. *Phys. Lett.*, B528:121–128, 2002.

[Levshakov04] S. A. Levshakov, M. Centurion, P. Molaro, and S. D'Odorico. Vlt/uves constraints on the cosmological variability of the fine-structure constant. 2004. [arXiv:astro-ph/0408188].

[Levshakov07.1] S. A. Levshakov et al. A new measure of delta alpha/alpha at redshift z = 1.84 from very high resolution spectra of q1101-264. 2007. [arXiv:astro-ph/0703042].

[Levshakov07.2] S. A. Levshakov, D. Reimers, M. G. Kozlov, S. G. Porsev, and P. Molaro. A new approach for testing variations of fundamental constants over cosmic epochs using FIR fine-structure lines. 2007. [arXiv:0712.2890].

[Linde86]   A. D. Linde. Eternally Existing Selfreproducing Chaotic Inflationary Universe. *Phys. Lett.*, B175:395–400, 1986.

[Lopez97]   R. E. Lopez, M. S. Turner, and G. Gyuk. Effect of finite nucleon mass on primordial nucleosynthesis. *Phys. Rev.*, D56:3191–3197, 1997.

[Lopez98]   R. E. Lopez and M. S. Turner. An accurate calculation of the big-bang prediction for the abundance of primordial helium. *Phys. Rev.*, D59:103502, 1999.

[Magg80]   M. Magg and C. Wetterich. Neutrino mass problem and gauge hierarchy. *Phys. Lett.*, B94:61, 1980.

[ManganoTalk07] G. Mangano. Relativistic energy density in the Universe. 2007. Plenary talk at Nuclear Physics in Astrophysics III, Dresden 2007.

[Marciano83] W. J. Marciano. Time Variation of the Fundamental 'Constants' and Kaluza- Klein Theories. *Phys. Rev. Lett.*, 52:489, 1984.

[Martins03] C. J. A. P. Martins et al. Wmap constraints on varying α and the promise of reionization. *Phys. Lett.*, B585:29–34, 2004.

[Melchiorri07] A. Melchiorri, L. Pagano, and S. Pandolfi. When Did Cosmic Acceleration Start ? *Phys. Rev.*, D76:041301, 2007.

[Minkowski77] P. Minkowski. mu $\to$ e gamma at a Rate of One Out of 1-Billion Muon Decays? *Phys. Lett.*, B67:421, 1977.

[Mota03]   D. F. Mota and J. D. Barrow. Local and Global Variations of The Fine Structure Constant. *Mon. Not. Roy. Astron. Soc.*, 349:291, 2004.

[MSW04] C. M. Muller, G. Schafer, and C. Wetterich. Nucleosynthesis and the variation of fundamental couplings. *Phys. Rev.*, D70:083504, 2004.

[Murphyprivate] M. T. Murphy. private communication. 2008.

[MurphyNH3] M. T. Murphy, V. V. Flambaum, S. Muller, and C. Henkel. Strong Limit on a Variable Proton-to-Electron Mass Ratio from Molecules in the Distant Universe. *Science*, 320:1611–1613, 2008.

[Murphy01] M. T. Murphy et al. Improved constraints on possible variation of physical constants from H I 21cm and molecular QSO absorption lines. *Mon. Not. Roy. Astron. Soc.*, 327:1244, 2001.

[Murphy03.1] M. T. Murphy, J. K. Webb, and V. V. Flambaum. Further evidence for a variable fine-structure constant from Keck/HIRES QSO absorption spectra. *Mon. Not. Roy. Astron. Soc.*, 345:609, 2003.

[Murphy03.2] M. T. Murphy et al. Constraining variations in the fine-structure constant, quark masses and the strong interaction. *Lect. Notes Phys.*, 648:131–150, 2004.

[NACRE99] C. Angulo et al. A compilation of charged-particle induced thermonuclear reaction rates. *Nucl. Phys.*, A656:3–183, 1999.

[NETGEN] M. Aikawa, M. Arnould, S. Goriely, A. Jorissen, and K. Takahashi. BRUSLIB and NETGEN: the Brussels nuclear reaction rate library and nuclear network generator for astrophysics. *Astron. Astrophys.*, 411:1195–1203, 2005.

[NillesSUSY] H. P. Nilles. Supersymmetry, Supergravity and Particle Physics. *Phys. Rept.*, 110:1–162, 1984.

[Nollett02] K. M. Nollett and R. E. Lopez. Primordial nucleosynthesis with a varying fine structure constant: An improved estimate. *Phys. Rev.*, D66:063507, 2002.

[Nordtvedt02] K. Nordtvedt. Space-time variation of physical constants and the equivalence principle. 2002. [arXiv:gr-qc/0212044].

[OliveSkillman04] K. A. Olive and E. D. Skillman. A Realistic Determination of the Error on the Primordial Helium Abundance: Steps Toward Non-Parametric Nebular Helium Abundances. *Astrophys. J.*, 617:29, 2004.

[Olive02] K. A. Olive et al. Constraints on the variations of the fundamental couplings. *Phys. Rev.*, D66:045022, 2002.

[Olive03] K. A. Olive et al. A re-examination of the Re-187 bound on the variation of fundamental couplings. *Phys. Rev.*, D69:027701, 2004.

[O'Meara06]   J. M. O'Meara et al. The Deuterium to Hydrogen Abundance Ratio Towards the QSO SDSS1558-0031. *Astrophys. J.*, 649:L61–L66, 2006.

[PDG06]   W. M. Yao et al. Review of particle physics. *J. Phys.*, G33:1–1232, 2006.

[PDG08]   C. Amsler et al. Review of particle physics. *Phys. Lett.*, B667:1, 2008.

[Peebles66]   P. J. E. Peebles. Primordial Helium Abundance and the Primordial Fireball. II. *Astrophys. J.*, 146:542, 1966.

[Peskin97]   M. E. Peskin. Beyond the standard model. 1997. [arXiv:hep-ph/9705479].

[Petrov05]   Yu. V. Petrov, A. I. Nazarov, M. S. Onegin, V. Yu. Petrov, and E. G. Sakhnovsky. Natural nuclear reactor oklo and variation of fundamental constants. i: Computation of neutronic of fresh core. *Phys. Rev.*, C74:064610, 2006.

[Pettini08]   M. Pettini, B. J. Zych, M. T. Murphy, A. Lewis, and C. C. Steidel. Deuterium Abundance in the Most Metal-Poor Damped Lyman alpha System: Converging on Omega_baryons. 2008. [arXiv:0805.0594].

[Pich95]   A. Pich. Chiral perturbation theory. *Rept. Prog. Phys.*, 58:563–610, 1995.

[Pieper01]   S. C. Pieper and R. B. Wiringa. Quantum Monte Carlo Calculations of Light Nuclei. *Ann. Rev. Nucl. Part. Sci.*, 51:53–90, 2001.

[Pudliner97]   B. S. Pudliner, V. R. Pandharipande, J. Carlson, S. C. Pieper, and R. B. Wiringa. Quantum Monte Carlo calculations of nuclei with A $\leq$ 7. *Phys. Rev.*, C56:1720–1750, 1997.

[RatraPeebles88]   B. Ratra and P. J. E. Peebles. Cosmological Consequences of a Rolling Homogeneous Scalar Field. *Phys. Rev.*, D37:3406, 1988.

[Reinhold06]   E. Reinhold et al. Indication of a cosmological variation of the proton - electron mass ratio based on laboratory measurement and reanalysis of h(2) spectra. *Phys. Rev. Lett.*, 96:151101, 2006.

[Rocha03]   G. Rocha et al. Measuring α in the Early Universe: CMB Polarization, Reionization and the Fisher Matrix Analysis. *Mon. Not. Roy. Astron. Soc.*, 352:20, 2004.

[Rosenband08]   T. Rosenband et al. Frequency ratio of al+ and hg+ single-ion optical clocks; metrology at the 17th decimal place. *Science*, 319(5871):1808–1812, 2008.

[Salam69]   A. Salam. Weak Interactions with Lepton-Hadron Symmetry. P. 367 of Elementary Particle Theory, ed. N. Svartholm (Almquist and Wiksells, Stockholm, 1969).

[Scherrer83] R. J. Scherrer. Primordial element production in universes with large lepton-baryon ratio. *Mon. Not. Roy. Astron. Soc.*, 205:683–690, 1983.

[Scherrer03] R. J. Scherrer. The uncertainty in Newton's constant and precision predictions of the primordial helium abundance. *Phys. Rev.*, D69:107302, 2004.

[Schlamminger07] S. Schlamminger, K. Y. Choi, T. A. Wagner, J. H. Gundlach, and E. G. Adelberger. Test of the Equivalence Principle Using a Rotating Torsion Balance. *Phys. Rev. Lett.*, 100:041101, 2008.

[SchwablStatMech] F. Schwabl. Statistische Mechanik. 2004. Springer Verlag, 2004 (2. Auflage).

[Serpico04] P. D. Serpico et al. Nuclear Reaction Network for Primordial Nucleosynthesis: a detailed analysis of rates, uncertainties and light nuclei yields. *JCAP*, 0412:010, 2004.

[Shaw05] D. J. Shaw and J. D. Barrow. Local experiments see cosmologically varying constants. *Phys. Lett.*, B639:596–599, 2006.

[Sisterna90] P. Sisterna and H. Vucetich. Time variation of fundamental constants: Bounds from geophysical and astronomical data. *Phys. Rev.*, D41:1034–1046, 1990.

[Smith92] M. S. Smith, L. H. Kawano, and R. A. Malaney. Experimental, computational, and observational analysis of primordial nucleosynthesis. *Astrophys. J. Suppl.*, 85:219–247, 1993.

[Steigman05] G. Steigman. Primordial nucleosynthesis: Successes and challenges. *Int. J. Mod. Phys.*, E15:1–36, 2006.

[Tegmark97] M. Tegmark. Is *the theory of everything* merely the ultimate ensemble theory? *Annals Phys.*, 270:1–51, 1998.

[Thorsett96] S. E. Thorsett. The Gravitational Constant, the Chandrasekhar Limit, and Neutron Star Masses. *Phys. Rev. Lett.*, 77:1432–1435, 1996.

[Tzanavaris06] P. Tzanavaris, M. T. Murphy, J. K. Webb, V. V. Flambaum, and S. J. Curran. Probing variations in fundamental constants with radio and optical quasar absorption-line observations. *Mon. Not. Roy. Astron. Soc.*, 374:634–646, 2007.

[Uzan02] J.-P. Uzan. The fundamental constants and their variation: Observational status and theoretical motivations. *Rev. Mod. Phys.*, 75:403, 2003.

[Vangioni-Flam02] E. Vangioni-Flam, K. A. Olive, B. D. Fields, and M. Casse. On the baryometric status of He3. *Astrophys. J.*, 585:611–616, 2003.

[Wagoner66] R. V. Wagoner, W. A. Fowler, and F. Hoyle. On the synthesis of elements at very high temperatures. *Astrophys. J.*, 148:3–49, 1967.

[Wagoner69]   R. V. Wagoner. Synthesis of the elements within objects exploding from very high temperatures. *Astrophys. J. Suppl.*, 18:247–295, 1969.

[Wagoner72]   R. V. Wagoner. Big bang nucleosynthesis revisited. *Astrophys. J.*, 179:343–360, 1973.

[Wegner72]   F. J. Wegner and A. Houghton. Renormalization group equation for critical phenomena. *Phys. Rev.*, A8:401–412, 1973.

[Weinberg67]   S. Weinberg. A Model of Leptons. *Phys. Rev. Lett.*, 19:1264–1266, 1967.

[WeinbergGRT]   S. Weinberg. Gravitation and cosmology: Principles and applications of the general theory of relativity. 1972. Wiley and Sons, 1972.

[WeinbergQFT2] S. Weinberg. The quantum theory of fields. vol. 2: Modern applications. Cambridge, UK: Univ. Pr., 1996.

[WeinbergQFT3] S. Weinberg. The quantum theory of fields. vol. 3: Supersymmetry. Cambridge, UK: Univ. Pr., 2000.

[Wendt08]   M. Wendt and D. Reimers. Variability of the proton-to-electron mass ratio on cosmological scales. 2008. [arXiv:0802.1160].

[Wetterich02.1] C. Wetterich. Probing quintessence with time variation of couplings. *JCAP*, 0310:002, 2003.

[Wetterich02.2] C. Wetterich. Conformal fixed point, cosmological constant and quintessence. *Phys. Rev. Lett.*, 90:231302, 2003.

[Wetterich03]   C. Wetterich. Crossover quintessence and cosmological history of fundamental 'constants'. *Phys. Lett.*, B561:10–16, 2003.

[Wetterich08]   C. Wetterich. Growing neutrinos and cosmological selection. *Phys. Lett.*, B655:201–208, 2007.

[Wetterich88.1] C. Wetterich. Cosmology and the Fate of Dilatation Symmetry. *Nucl. Phys.*, B302:668, 1988.

[Wetterich88.2] C. Wetterich. Cosmologies with variable Newton's 'constant'. *Nucl. Phys.*, B302:645, 1988.

[Williams04]   J. G. Williams, S. G. Turyshev, and D. H. Boggs. Progress in lunar laser ranging tests of relativistic gravity. *Phys. Rev. Lett.*, 93:261101, 2004.

[Wilson71]   K. G. Wilson. Renormalization group and critical phenomena. 1. Renormalization group and the Kadanoff scaling picture. *Phys. Rev.*, B4:3174–3183, 1971.

[Wilson73]   K. G. Wilson and J. B. Kogut. The Renormalization group and the epsilon expansion. *Phys. Rept.*, 12:75–200, 1974.

[WMAP3]　　　　　D. N. Spergel et al. Wilkinson Microwave Anisotropy Probe (WMAP) three year results: Implications for cosmology. *Astrophys. J. Suppl.*, 170:377, 2007.

[WMAP5]　　　　　E. Komatsu et al. Five-Year Wilkinson Microwave Anisotropy Probe (WMAP) Observations:Cosmological Interpretation. 2008. [arXiv:0803.0547].

[YooScherrer02]　J. J. Yoo and R. J. Scherrer. Big bang nucleosynthesis and cosmic microwave background constraints on the time variation of the Higgs vacuum expectation value. *Phys. Rev.*, D67:043517, 2003.

[Zahn02]　　　　　O. Zahn and M. Zaldarriaga. Probing the Friedmann equation during recombination with future CMB experiments. *Phys. Rev.*, D67:063002, 2003.

VDM Verlagsservicegesellschaft mbH

Die VDM Verlagsservicegesellschaft sucht für wissenschaftliche Verlage abgeschlossene und herausragende

Dissertationen, Habilitationen, Diplomarbeiten, Master Theses, Magisterarbeiten usw.

für die kostenlose Publikation als Fachbuch.

Sie verfügen über eine Arbeit zu aktuellen Fragestellungen aus den genannten Fachgebieten, die hohen inhaltlichen und formalen Ansprüchen genügt, und haben Interesse an einer honorarvergüteten Publikation?

Dann senden Sie bitte erste Informationen über sich und Ihre Arbeit per Email an *info@vdm-vsg.de*.

Sie erhalten kurzfristig unser Feedback!

VDM Verlagsservicegesellschaft mbH
Dudweiler Landstr. 99   Telefon +49 681 3720 174
D - 66123 Saarbrücken   Fax   +49 681 3720 1749
www.vdm-vsg.de

Die VDM Verlagsservicegesellschaft mbH vertritt

Printed by Books on Demand GmbH, Norderstedt / Germany